TO [illegible] or not
TO BELIEVE

To Nagarjuna,

Thanks for sharing this Mystery we call our lives...

www.RahasyaPoe.com

TO BELIEVE *or not* TO BELIEVE

The Social and Neurological Consequences of Belief Systems

RAHASYA POE

Library of Congress Control Number:		2009910665
ISBN:	Hardcover	978-1-4415-8728-2
	Softcover	978-1-4415-8727-5

This book was printed in the United States of America.

To order additional copies of this book, contact:
Xlibris Corporation
1-888-795-4274
www.Xlibris.com
Orders@Xlibris.com

62908

CONTENTS

Part II: The Neurological Consequences

Part III: Rewriting Our Human History

TESTIMONIES AND COMMENTS

Rahasya's detailed exploration of "belief" and the various ways in which it keeps us locked in destructive modes of thinking and experiencing life is a wonderful addition to the modern spiritual library. Reading the book itself is an interesting exercise in watching the way "beliefs" operate in our lives. He explores a wide range of themes from an even wider collection of viewpoints. As the Buddha said, "Find out for yourself what is true."

Bill Martin, *The Parent's Tao Te Ching* (as seen on Oprah's Book Club)

"Why do we believe what we believe? Poe asks this critical question at a very critical time in history and delves deeply into a mind-boggling examination of the neurology of belief and the sociological consequences and implications for our future. The book covers the spiritual, scientific, and psychological origins of belief, including the influence of the world's great religions, with input from scientists, researchers, and spiritual leaders on how we can change those beliefs to empower ourselves, humanity, and the planet. The author devotes ample time to the neuroscience behind belief and comes to the stunning realization that our beliefs do indeed affect our brains, often with negative results. Poe's book is a comprehensive and enlightening look at what we have come to believe, personally and collectively, and why we must challenge the paradigms that no longer serve us."

Marie D. Jones, author of *2013: End of Days or A New Beginning? —Envisioning the World After the Events of 2012*

We will have won this war of ideas against religion when atheism is scarcely intelligible as a concept. We will simply find ourselves in a world in which people cease to praise one another for pretending to know things they do not know. This is certainly a future worth fighting for. It may be the only future compatible with our long-term survival as a species. But the only path between now and then, that I can see, is for us to be rigorously honest in the present. It seems to me that intellectual honesty is now, and will always be, deeper and more durable, and more easily spread, than "atheism."

Sam Harris, author of *Letters to a Christian Nation* and *End of Faith*

How do you get out of a belief system? First you have to destruct the belief system. Traditionally, the teacher is supposed to remove your ignorance. But when you remove ignorance, you start with removing what is causing the ignorance, which is your belief system. So the teacher's job indeed is to first deconstruct your belief system. And then to give you inspiration so you'll go out to create a path to discover what is spirit, what is beauty, what is love, because these things nobody can teach you. So teaching really should be a demolition job.

Amit Goswami

Fundamentalism, in and of itself, is benign and can be personally beneficial, but the anger and prejudice generated by extreme beliefs can permanently damage your brain.

Dr. Andrew Newberg, author of *How God Changes Your Brain*

The first thing we need to do is realize that some of our old beliefs are dysfunctional and no longer work, if indeed they ever did. If the desire to do that isn't there, there's nothing to do.

Fred Alan Wolf, author of *Taking the Quantum Leap*

The difficulty in today's world is our technology and science has outrun our theological advances. The reason for that is in technology and science, we have had the courage to ask the single question that theology has been afraid to ask, Is it possible that there's something I don't know about this, the knowing of which would change everything?

~~~~

What if everything we thought about God up to this point was not necessarily true? Tomorrow's God is what emerges from our willingness to question our prior assumption.

Neale Donald Walsch, author of *Conversations with God*

What the brain likes to do is to simply replace old ideas or beliefs with new ones.

Dr. Joe Dispenza, author of *Evolve Your Brain*

I see the main problem as a spiritual one, not a resource problem, or a problem with this or that government, but a larger problem centered around human beliefs, the troublesome elements founded in our mythology. Our problematic mythology is collapsing all around us. It is a mythology that is predatory.

Albert Villoldo, PhD
~~~~

ACKNOWLEDGMENTS

I would like to thank you, the reader, for having the courage and taking the time to explore the possible social and neurological consequences of believing without evidence.

I would also like to thank the many people I have interviewed and talked with who have been speaking out for some time from their own perspectives on the subject of science, religion, and belief systems. This is no small thing when we are confronted by a world that is deeply embedded with superstition and beliefs that date back thousands of years to when we knew very little about science, evolution, archaeology, or the cosmological truths we have uncovered during the past century.

I would also like to thank my many friends and family members for the encouragement that it takes to write a book such as this. Among them are my editor, Karen Bleske; my webmaster, Kasey McKnight; my friend and auditor, Mike "Misha" Hendrickson, who gave so much good advice on the first manuscript; my very good friend Tanmayo, who has helped in countless ways with her feedback and support; my dad, who has been a living example that it is never too late to change and evolve; my daughter Cindy, whom I love up to the sky and past it; my grandson Jason, who lives in the world that we all need to change; my son Jason, who constantly reminds me in his own way of how much I've changed; and my friend Mona Lewis for her ability to be objective. I want to thank Pamela Jamian for her honest feedback and assistance. I also want to thank Timothy Freke, who brought a refreshing sense of humor to this subject.

And last, but by no means least, my wife Dhara Lemos, who has been a constant source of encouragement and inspiration. She and Craig Perry are also responsible for the stunning cover design.

Wm "Rahasya" Poe
www.RahasyaPoe.com

To Believe or Not To Believe:
The Neurological and Social Consequences of Belief Systems

Those who can make you believe absurdities can make you commit atrocities.
Voltaire (1694-1778)

Dedicated to my children, Cindy and Jason, my grandson Jason, my stepson Siddhartha, and all the other future generations of children who are going to look back at our generation and the legacy we are leaving them and who will ask, "What did you do to help make the world a safer and better place to live?"

WARNING: This book may change your worldview!

Now, an enormous and unprecedented transformation in the human experience is underway. Our perception is expanding beyond the limitations of the five senses. We are becoming aware of ourselves as more than bodies and minds, molecules and enzymes. We are becoming aware of our lives as purposeful and our experiences as meaningful. Our focus is shifting from exploration of the outside world to the exploration of interior experiences and their relationships with the outside world. This is the new frontier of human inquiry, and all seekers of truth are being drawn to it.
Gary Zukav, *Soul Stories*

PREFACE

In light of the obvious fact that on every level our world is not functioning well, I think we need to start looking for and finding the common denominator underlying world events and the decisions being made in every area of society. When we look at world events as shown to us in the media, things seem chaotic and disturbingly complex. But as we go below the superficial layers of society we find one simple common denominator that dictates most of our actions . . . our beliefs.

Even though a good part of this book is about "questioning some of our answers," a few questions need to be asked in light of new information. One is this: Since roughly half the earth's population believe in some form of the Judeo-Christian-Muslim religions and organize their entire lives, nations, and armies around them with the assumption that they are true, what if we now have hard evidence that the original premise for all three religions is false? If there is such evidence, don't we owe it to ourselves and future generations to research it thoroughly? The answer to this question might also explain why no one can agree, even within each religion. Truth has always and will always stand on its own; however, a lie must always be supported and constantly held up by other lies. This seems to be the situation we find ourselves in, and it has led to a continual stream of *social consequences* on our planet that now threaten our very survival if we don't find a remedy soon. The remedy could lie in our ability to redefine ourselves as human beings and to do this will require nothing less than rewriting our entire human history.

As we explore the deeper implications of belief systems and limiting beliefs in general, we also run into the *neurological consequences* of maintaining extreme beliefs in a world that is desperately attempting to evolve into a rational and sustainable planet. As it turns out, the evidence clearly demonstrates that there are consequences on a neurological level when a person is deeply embedded in beliefs that are illogical and even counterintuitive. In the words of Dr. Andrew Newberg, *"The anger and prejudice generated by extreme beliefs can permanently damage your brain."*

This statement must be taken seriously in light of the research that is being done using advanced fMRI technology.[1]

And toward the end of this book, we will take a step beyond what most people could accept as even a possibility. I ask the question: What if we have been visited by an advanced race in our past? An extreme question? Yes. But when you explore the evidence, you realize there is much more to our history than that we have been told to consider. I am also surprised at how many people simply don't read and therefore do not know about the discoveries in the last century. Surprisingly, not only scientists and researchers validated this new information, but biblical scholars and historians have also. We will look at why the Catholic Church is starting to position itself for the time this information is released so it will be ready for the worldwide religious repercussions the knowledge will cause.

I noticed while writing this book that my research inevitably gave rise to an anger within myself that I was not aware of, so in a sense this has been my journey to bring these emotions to the surface and to process them through a deeper understanding that arises when we have better information. I hope you will be able to do the same, because if we continue to do business as usual we will find ourselves deeper and deeper in the chaos and madness that surrounds us in today's world. But there is simply no way around it; we must look honestly at our beliefs that are outdated and absurd in light of today's science and understanding of the world we live in. The only way to do this is to search them out, in ourselves first, and then in others, and to take a hard cold look at them in the light of day. Then and only then will we be able to create the silent space inside ourselves that we need to make appropriate decisions that will make positive and lasting changes in our world in a sustainable way—sustainable being the key word.

I just want to say a quick word here on sustainability and global warming. There's probably no better example of the danger of our tendency to "believe

---

[1] Functional MRI or functional magnetic resonance imaging (fMRI) is a type of specialized MRI scan. It measures the haemodynamic response related to neural activity in the brain or spinal cord of humans or other animals. It is one of the most recently developed forms of neuroimaging. Since the early 1990s, fMRI has come to dominate the brain mapping field because of its low invasiveness, lack of radiation exposure, and relatively wide availability. From www.wikipedia.com.

without evidence" than the global warming issue. I have yet to meet one person who has said, "Based on the scientific evidence I don't think global warming is true." What I have heard is "I don't 'believe' in global warming." Most rational and aware people would agree that global warming is not something to believe in, or not, as if it's a religion, yet we have millions of people who simply "believe" it's a false concept in the face of overwhelming evidence. The suggestion here is that this is a cognitive process set in motion by preexisting neural pathways caused by early indoctrination into belief systems.

We all have a human psychological tendency that we must also address before you start reading this, or any other, book that may conflict with your personal beliefs. Besides the fact that it is very difficult to expose and dismantle previously held beliefs, something that psychologists have been studying for many years is our "denial of death" and how to some degree it influences almost every decision we make. Because we deny death, we tend to filter and block out a lot of information that brings us face-to-face with that reality. Ernest Becker came to the realization that psychological enquiry inevitably comes to an end at some point and we hit a wall, which is our limits of understanding, and when this happens it evokes our tendency to create beliefs to satisfy and placate our minds; in other words, we create beliefs to keep us unconscious so we can continue the dream of being awake and "believing" we know what's going on in our world.[2] The idea of what

---

[2] According to Wikipedia, Dr. Ernest Becker was a cultural anthropologist and interdisciplinary scientific thinker and writer. Becker came to the realization that psychological inquiry inevitably comes to a dead-end, beyond which belief systems must be invoked to satisfy the human psyche. The reach of such a perspective consequently encompasses science and religion, and he became known as the creator of a "science of evil" because it puts into perspective the reasons why we create religious belief systems. Because of his breadth of vision and avoidance of social-science pigeonholes (given the independence of his thinking in the 1960s), Becker was awarded the Pulitzer Prize in 1974 for his 1973 book, *The Denial of Death.* In the past two decades, a trio of experimental social psychologists have amassed a large body of empirical evidence substantiating the universal motive of death denial as advanced by Becker. The highly topical and jargon-free account of that work is now in print *In the Wake of 9/11: The Psychology of Terror* by Thomas A. Pyszczynski, Sheldon Solomon, and Jeff Greenberg. (American Psychological Association

happens to us after death is one such wall. More recently, neuroscientists have also discovered with the help of fMRIs that our brains have a strong tendency to rely on preexisting beliefs to make decisions instead of taking the time to "think out of the box" using logic and new information.

So at some point we may run into difficulty communicating as we start disassembling beliefs through deeper understanding and new information, because this process will bring us once again face-to-face with what we have been running from and what haunts us on a very deep level . . . our mortality and the unknown. Keep in mind that we have never in the history of the world witnessed a resurrection; believing in resurrection is one thing, experience and knowing is another. Studies have shown that religion is a direct outcome of our attempt to deal with death, and more to the point, our denial of death. We create religions, ideologies, and nations as an extension of ourselves in an attempt to gain immortality through their continuation. Because we identify with them we are willing to die for these same creations.

All of the information you will read here is factual and I've made every attempt to keep judgment out of it. When judgment does occasionally creep in, I try to make it clear that this is my point of view and I reference what that view is based on. I do not want you to believe or disbelieve me because both mental processes are the same and will keep you from thinking for yourself. And this should be the goal of every human being—to think for himself or herself. The mere act of believing is our greatest obstacle to knowing. The assertion of this book is that we may be evolving to a level of consciousness where we no longer need to go around pretending we know things we have no way of knowing, which is what we call our "beliefs." Is it not possible that we are finally growing up spiritually and at last are able to say, "I don't know," and then set out to explore this mystery called life without dragging beliefs, superstitions, and outdated concepts behind us? We are still on the shoreline and just ankle deep in this mysterious ocean of life.

We turn toward the truth under the same heliotropic laws as a flower turns toward the sun, and for the same reason . . . connection with the source of energy.

Rahasya Poe

Press, 2003). There's also a great DVD out called *Flight from Death: Quest for Immortality* (2003).

Another issue also must be addressed and it has to do with our nationalistic beliefs. We are at a crossroads in the United States of America and I'm not at all sure we are going to make it unless we wake up to what's blinding us. We have become the laughingstock of most of the world, having one of the worst and most dangerous health-care systems in the developed world, an educational system that is designed to fail with underpaid teachers, a political system that can call anyone a military combatant and lock him up for three years with no due process, a private paramilitary combat force that received a billion-dollar contract under the Bush administration to start putting private military forces on our streets, and the list actually gets worse from this point on.[3] [4] What's important in reference to the topic of this book is the fact that these events would not be possible except for the fact that they are done behind the veil of illusion that keeps people divided from the truth. The tools to accomplish this are getting people to "believe" lies supported entirely by faith, the faith in the belief that this is "your" government and it represents your interests. I'm sorry to break it to you but this is no longer your government, and with a weakened democracy it's going to be difficult to get it back from those in charge. How do we get it back? Read this book for one, so you can start thinking for yourself again and stop being dependent on the beliefs that have been handed down to you and injected into you through the corporate media. It's no wonder that we are controlled by corporate interests because, in fact, our states are incorporated along with the United States; we live under corporate rules and regulations.

That being said, I also feel compelled to say that with the recent election of Barack Obama, I have a renewed outlook for the possibilities ahead. This is a good indication that a lot of people are waking up and seeing through the illusion that most politicians hide behind. But even still, we must stay vigilant and demand transparency in our government. Even now, a few

---

3 I'm referring to Blackwater. These forces have been brought up on numerous charges for drugs, misuse of funds, and shooting innocent civilians in Iraq. Consequently there have been no major repercussions from these events.

4 Read *The End of America* (2009) by Naomi Wolf. She is a journalist and past adviser to Al Gore and she is now on "The List," making it almost impossible to travel and not be harassed. This is entirely predictable as a democracy turns into a dictatorship.

months after the election, I can see a difference between his campaign words and his presidency. Carl Sagan once said:

> *Fundamental changes in society are sometimes labeled impractical or contrary to human nature: as if nuclear war were practical or as if there were only one human nature. But fundamental changes can clearly be made. We are surrounded by them. In the last two centuries abject slavery, which was with us for thousands of years, has almost entirely been eliminated in a stirring worldwide revolution. Women, systematically mistreated for millennia, are gradually gaining the political and economic power traditionally denied to them. And some wars of aggression have recently been stopped or curtailed because of a revulsion felt by the people in the aggressor nations. The old appeals to racial, sexual and religious chauvinism and to rabid nationalism are beginning not to work. A new consciousness is developing which sees the earth as a single organism and recognizes that an organism at war with itself is doomed. We are one planet.*

The Purpose and Goal of This Book

The purpose and goal of this book is to show through evidence that with the practice and maintaining of rigid beliefs, we set up deeply entrenched neural networks that not only reflect the limitations of our beliefs but also limit and distort the way we see the world around us. We have many terms for such people, terms such as "tunnel vision," "hardheaded," "shortsighted," and a long list of psychiatric labels to point out that the person in question simply cannot understand what another person may see clearly. One of the scientific terms is "attentional blindness," and we will hear what the lead researcher of HeartMath, Rollin McCraty, has to say about this later. I see this as something to be concerned with in today's world because it suggests that we may need to let go of our limiting beliefs to envision a new reality.

It's no secret, to anyone who cares to know, that the profit from weapons sold to Third World countries by the United Kingdom and the United States is far more than the aid they give those same countries. For far too long First World countries have worked behind a veil that is possible only because "we the people" were not thinking for ourselves. In light of massive amounts of documentation, it has become absurd to even pretend that our intentions are honorable and just. As we break down hardened beliefs we will open up new ways of seeing the world, ways in which we can cocreate a better world. This can happen only if we are honest with ourselves, and herein lies the challenge.

It is also the purpose of this book to point out that it may be time to rewrite our human history. For the last few centuries we have accepted a biblical history that has been eroding because of science and the human passion for understanding. You will find within these pages what I believe will be the final blow to organized religions. Even now, biblical scholars and even the Vatican are recognizing the overwhelming evidence that all is not well with our biblical interpretation of history, that along with the fact that it simply doesn't work in a modern-day society with high-technology. In fact, if anyone were to try to live out in today's world the laws laid down in scripture, he or she would end up in prison or labeled a terrorist. But history has shown that apathy and silence have aided in the expansion of superstition for far too long. Later, we will let the evidence speak for itself, because remember this: The whole point is for you to think for yourself, do your own research, and come to your own conclusions based on the best evidence.

To create a new and better future, we must first see where we are and how we got here so we don't end up dragging fragments of our past into the future, fragments that end up being the seeds of our destruction. We must start fresh, clean, and with a clear vision of what needs to be done and this will come about only after we expose our faulty and limited way of dealing with life, which is the way we believe life has to be. If we want to evolve and move on we must first release ourselves from our primitive past beliefs and superstitions, which is what this book is about—creating a new world by identifying and letting go of our outdated and dysfunctional beliefs.

Part I:
The Social Consequences

CHAPTER 1

Through the Eyes of a Child

I would like to share with you the experience that started me down this path. It was the first of many such experiences in my life that nudged me in small degrees in another direction. It's usually later in life that we realize their full impact. I might also add that it was one of many such experiences that helped me to realize the importance of the acceptance of another point of view or belief. Acceptance doesn't necessarily mean you agree. Without the ability to see clearly and to embrace another person's belief without judgment or condemnation, we are forever locked in conflict, not only with them, but with ourselves. But the first thing to do, and the most difficult, is to look at our own beliefs with the courage of a spiritual warrior. We must be willing to pluck the log from our own eye before we start on the "others."[5]

I was eight years old and growing up in West "by God" Virginia. Life was simple during the '50s. We still had horse-drawn ice wagons that delivered ice to homes with iceboxes and a man who rode around on a cart pulled by a donkey; he sold rags and sharpened everything from scissors to knives. I had an uneventful childhood for the most part. I had a dog and my Dad bought me my first shotgun when I was eight years old. I played hide-and-seek down by the river and it was still safe to go down by the train tracks and listen to the hobos talk of their experiences. My media entertainment in those days was listening to *Amos and Andy, The Shadow, The Jack Benny Show,*

---

[5] Luke 6:42: . . . How can you say to your brother, "Brother, let me remove the speck that is in your eye," when you yourself do not see the plank that is in your own eye? Hypocrite! First remove the plank from your own eye, and then you will see clearly to remove the speck that is in your brother's eye. (King James version)

and *The Lone Ranger* on the radio. I always looked forward to walking to and from church every Sunday with my mother and father; besides, there was not much else to do as our town was barely on the map. Even at that early age I was constantly asking those pesky questions in Sunday school, questions such as, "Where did Cain and Abel get their wives if they were the only two children of Adam and Eve?" And when the Bible said that the sons of God saw the daughters of men to be fair and took them to be wives whomever they chose, it naturally made me ask, "Who are the sons of God?"[6] I remember the Sunday school teacher telling my Dad that I was probably too young to bring to the class. But it was a fire and brimstone Bible-thumping preacher on a hot summer morning who brought me to my senses, although I don't think that was his intention.

Just think of the tragedy of teaching children not to doubt.
Clarence Darrow (1857-1938)

I always listened to the preacher but usually didn't understand much at that age. He came across very aggressively so I usually sat there fanning myself with a little cardboard fan that had a picture of Jesus on one side and an advertisement for the local funeral parlor on the other. I can also remember looking out through the part of the stained-glass window that you could see through and wishing I were outside playing, occasionally wondering why the preacher was so angry and upset. But one Sunday morning he started talking about "another" religion, which totally rocked my little world because it surprised me to learn there was, or could be, any such thing as *another* religion. How could that possibly be? How could there be another God? It must have been the Hindu religion he was talking about because I remember his mentioning how these people were going to go to Hell because they worshipped cows and believed they were sacred. Well, this seemed quite strange to me and really got me to wondering how these people could believe that was true, especially since we regularly grind cows up and eat them almost every Sunday right after church. Something had to be wrong with this picture.

---

6 Genesis 6:2: That the sons of God saw the daughters of men that they were fair; and they took them wives of all which they chose. 4: There were giants in the earth in those days; and also after that, when the sons of God came in unto the daughters of men, and they bore children to them, the same became mighty men which were of old, men of renown. (King James version)

On the way home I asked my Dad how is it that we know that our religion is true and theirs is not. He said, "Well, because we *believe* it is and we have *faith* that what we believe in is true, and in fact, believing it is what makes it true." I remember walking for a while, the wheels in my head turning, and then asking the next, for me, obvious question: "If we were born where they were born, wouldn't we believe that we were right to worship cows and wouldn't we have faith that what we believed was true?" He said, "Well, yes, we would believe we were right." *I could hear the uncertainty in his voice.* Then I said, "But they're wrong, aren't they?" My Dad said, "Yes, they're wrong." Of course the insanity of this type of reasoning becomes apparent even to an eight-year-old because we are quick to invalidate someone else's belief for the very same reason we use to validate our own. Looking back, I find it interesting that I believed in Santa Claus a year longer than I believed in there being only one true God, so not being able to believe something wasn't the problem. I stopped believing in Santa Claus when I started realizing the impossibility of his going to so many homes in one night, and he would never fit down our fireplace chimney anyway, not to mention the fact that reindeer don't fly. I also learned how easy it was for the adults in my life to avoid answering questions such as this from an eight-year-old boy . . . but I'm older now and I'm still asking, so here we go.

Toltec information about the most important principles to live by, from *The Four Agreements* (1997) by Don Miguel Ruiz, who also said,
"Don't believe anything."

1. Be impeccable with your words.
2. Don't take anything personally.
3. Don't make assumptions.
4. Always do your best.

CHAPTER 2

Making Sure We Are on the Same Page

It pays to just touch base and make sure we are still together in our little adventure because, as I mentioned before, we all have an ego-defense system that filters out any new information that conflicts with old beliefs. If you must believe something, believe this: If you let that happen (filter out what doesn't fit your beliefs), you will find yourself divided as to this information I am presenting because this is a strong indictment of belief systems backed by a lot of evidence, both historical and scientific, and once you're divided, your ego/personality has you right where it wants you, in conflict. Once you are in inner conflict, you are caught in another trap and would probably just end up creating another belief system that would rob you of what little consciousness you may have left. And remember, I'm not out to convince you, I don't want you to believe me without evidence; I want you to come to your own conclusions by looking at the evidence. I'm not claiming I believe this because it makes me feel good or gives my life meaning. I believe this only because it is standing the test of time and is open to further inquiry. It's very important to realize that to believe without evidence or further inquiry, even if what you believe happens to be true, will not initiate a transformation of consciousness or awareness. Let me repeat that:

It's very important to realize that to believe without evidence or further inquiry, even if what you believe happens to be true, will not initiate a transformation in consciousness or awareness.

Rahasya Poe

Definitions

I suppose it would be prudent, if not wise, to define some of the terms I will be using in writing this book. We all have a tendency to get lost in words that have picked up baggage from one generation to another, not to mention the intentional manipulation of truth by slightly changing the meaning of a word, and the numerous translations in which words were not translated properly because the new language did not allow for particular terms or meanings. It even gets worse when those words are in what we call Holy Books that are supposed to be the inspired and immutable written truth of the creator of our universe.[7] So let's take a look at some of the important words we'll be using.

Belief:

Men willingly believe what they wish.
Julius Caesar (100-44 BC)

Something believed; an opinion or conviction: a belief that the earth is flat.[8] *A state or habit of mind in which trust or confidence is placed in some person or thing.*[9]

When we stop and really analyze this definition, we can see a problem arising right away because many of us forget that most, if not all, of our beliefs are just thoughts that we process over and over until they became a mental habit, not to mention that the thought was most probably given to us by some power figure in our lives and we accepted it on faith . . . faith that the person who told us had thought it out, researched it at least some, and came to this conclusion through some type of verification or deduction. What we often do not know is that the belief came to that person under the same system of blind acceptance, generation after generation.

---

7 There are literally thousands of known changes in the Bible, some with good intentions and some with bad intentions. Either way proves that the Bible is not the inspired words of the creator of the universe, unless of course he is a God who simply can't make up his mind. See *Misquoting Jesus* (2007) by Bart D. Ehrman.

8 Based on the Random House Unabridged Dictionary, 2006.

9 *Merriam-Webster's Collegiate Dictionary*, 2006.

I can believe there are things I cannot see without believing "in" things I cannot see.
Rahasya Poe

When I use the word "belief" in this book I am referring to the kind of rigid belief you find in religions in the form of dogma and long-accepted traditions whose authority is no longer questioned. For instance, I have a belief that the sun will rise tomorrow based on the fact that it has done so every morning that I can remember, at the same time always keeping in mind that I don't "know" the sun will rise because that lies in the future. Believing anything on blind faith is one thing, but coming to accept it as unquestionable truth, part of a holy scripture written by the creator of the universe, and a truth whose validity can no longer be questioned is quite another. So it is when "belief" slips over into "pretending to know—and forgetting that we're pretending" that we get ourselves into trouble and end up spreading that trouble to "others," others who we interestingly called "unbelievers" for the simple reason that they do not believe what we believe. If you are feeling the slightest little bit uneasy now, it's probably because we have touched upon a belief you have that was part of your indoctrination as a child and has now become part of your identity.

In other words, when we have a belief, we are pretending (consciously or unconsciously) that we know something that in fact we do not know and probably cannot know. In psychiatric circles this is called a delusion and can lead to a psychotic break with reality if the belief cannot be reconciled with the accepted reality around the person in question.

And as I mentioned before, "beliefs" as used in this book also refers to those beliefs that are limiting and divisive, including self-limiting beliefs, nationalism, ideologies, religions, and so on. I wasn't at all surprised to find that some of the definitions for "delusion" were the same as belief; it was just a matter of degree. A good rule of thumb is if you can easily substitute the word "intention" for belief or believe then it's probably not the definition we are discussing in this book.

Faith:

Faith must have adequate evidence, else it is mere superstition.
Alexander Hodge (1823-1886)

Firm belief in something for which there is no proof.[10]

Again, we have a word that means many things to many people. I'm not going to be using the word "faith" as it might be used to have faith in your abilities, or to have faith that the sun will rise tomorrow, or to have faith in yourself. I'm not sure we could live without this type of faith. The faith that gets us in trouble is when we accept beliefs entirely on faith with no evidence, and in fact one of the requirements of a "faith-based belief" is that there can be no evidence. Faith can lead us through the darkness of the unknown but it can also blind us to the facts. "*Now faith is the substance of things hoped for, the evidence of things not seen.*"[11] What we need is some rational thinking along the way so we need not throw out the baby with the bathwater when dealing with these terms, but we do need a new level of awareness and understanding so we don't find ourselves in a situation like that of the Vatican. In 1634, at the age of 70, Galileo was brought before the Vatican inquisition and threatened with torture if he did not publicly retract his view that the earth orbits around the sun. He was sentenced to imprisonment and spent the rest of his life under house arrest. The persecution of Galileo remained an embarrassment for the Catholic Church for the next three centuries. The Vatican finally forgave him in 1992. Blind faith and arrogance made it wait three centuries to admit what everyone else on the planet already knew.

Now we find ourselves with beliefs based entirely on faith and we no longer have the luxury of waiting three centuries to admit we could be wrong. Since I am writing this in a predominately Christian country, it will be easy to see the preposterousness and danger of thinking that if I strap on explosives and go blow up some nonbelievers (people who don't believe what I believe) that I will immediately go to heaven, have 72 virgins for eternity, and I will also get to take some of my family with me. Now that

---

10 *Merriam-Webster's Collegiate Dictionary*, 2006.

11 Hebrew 11. (King James version)

we've entered the atomic age, this belief becomes even more problematic because there are thousands of eager fundamentalist Muslims waiting in line for this privilege; even their mothers applaud them in admiration and they are found to be smiling when they push the button. It doesn't take much to see that faith in a belief such as this can drive a person to commit unheard-of atrocities because he was made to believe absurdities. It is also probably appropriate to include the definition of the word "delusion."

"Delusion:" *A fixed false belief that is resistant to reason or confrontation with actual fact.*[12]

The reason I started this book with the quote from Voltaire, *"Those who can make you believe absurdities can make you commit atrocities,"* is that this is what this book is all about. The neural networks that make believing absurdities a reality are the same networks that allow a person to be manipulated and controlled. To believe something absurd requires a person to essentially limit his or her understanding of what others would see as logical and rational and indeed, obvious. It will be important for you to understand the link between "believing absurdities" and "committing atrocities" as we continue. This point is something that has been overlooked until recently with the advances of fMRI technology.

In the same way that I cannot jump up and startle myself, I cannot believe something without the conviction that it is true. This conviction can come about in two ways: One would be through a rational approach to proving or disproving. The other is through "faith," which is the only way religions have found to circumvent reason without actually needing to address the facts; you either believe or you don't—period. To question your belief means that you have failed and you are said to have "lost your faith." Of course it would be difficult if you were the only one doing these mental gymnastics, so finding others and convincing them through whatever means to have faith that what you are believing is true is instrumental in creating the illusion of truth and sustaining it generation after generation. If this doesn't work, you can always set out to eliminate the nonbelievers.

Scientists also have faith, although they would hesitate to use this word for fear of its being misunderstood. They would be more likely to use

[12] From Dictionary.com.

the term "adventurous spirit" or simply "curiosity." They have faith in a theory's being correct, something that theoretical physicists know well. This is different from religious faith in that scientists start immediately to find proof and validation for that in which they have faith. They are also open to controversy (or at least should be) and will change or drop their theories if they do not prove to be true after further research and evidence. But even some of the best of scientists have been known to take on a "religious attitude" with their work and hold on to their theories long after they were shown to be incorrect.

Now, I hear some of you saying, "But what about the faith that people such as Mother Teresa had?" Well, let's look at someone we were told had unwavering faith. Recently a letter she wrote in confession to Jesus was released in the book *Mother Teresa: Come Be My Light* and also in an article in *Time* magazine, August 23, 2007, and here it is:

> *Lord, my God, who am I that You should forsake me? The Child of your Love—and now become as the most hated one—the one—You have thrown away as unwanted—unloved. I call, I cling, I want—and there is no One to answer—no One on Whom I can cling—no, No One.—Alone . . . Where is my Faith—even deep down right in there is nothing, but emptiness & darkness—My God—how painful is this unknown pain—I have no Faith—I dare not utter the words & thoughts that crowd in my heart—& make me suffer untold agony.*
>
> *So many unanswered questions live within me afraid to uncover them—because of the blasphemy—If there be God—please forgive me—When I try to raise my thoughts to Heaven—there is such convicting emptiness that those very thoughts return like sharp knives & hurt my very soul.—I am told God loves me—and yet the reality of darkness & coldness & emptiness is so great that nothing touches my soul. Did I make a mistake in surrendering blindly to the Call of the Sacred Heart?*

At the end it was addressed to Jesus and was undated. This stands as a testimony, along with many other letters, that in our deepest most truthful inner moments we all understand that we live in a mystery, for which there are few answers when we drop the beliefs we made up to subdue our fears. It's obvious from this letter that Teresa has many doubts about the path she chose, yet the church says this is a testament to her "unwavering faith" and interprets it as showing God's grace. Maybe I missed something in the letter but I read, "*I have no faith.*" And if this were the only instance it could be an isolated moment of depression, but in March 1953, she wrote Archbishop Périer, *"Please pray specially for me that I may not spoil His work and that Our Lord may show Himself—for there is such terrible darkness within me, as if everything was dead. It has been like this more or less from the time I started 'the work.'"* How is it that an organization of well-educated people can interpret this type of extreme doubt as proving their doctrine? One answer, of course, could be their limited neurological connections blinding them to the personal needs of Mother Teresa in favor of protecting the image of the Church. I feel sorry for Mother Teresa because it seems obvious that she was reaching out for help in many directions, but she became such a symbol for the Church that her requests were all but ignored and covered up. I know some of you probably *"believe"* that you shouldn't speak this way of the dead, but if we need to believe something in reference to a person such as this, let's believe that we can still help her by airing out her writings and feeling her needs with the compassion she should have received while alive. Besides, if she didn't intend that her writings be read, she would have made sure to destroy them or have left them in such a way that they would never have been made public.

Someone else worth looking into on the issue of faith is Emanuel Swedenborg, who spoke out strongly against depending on faith as a way to salvation and heaven.[13] To use faith in such a way that it lights our path toward knowledge is clearly valuable; however, to use it as a foundation to validate truth is precarious and misleading at best.

One reason I find myself criticizing religion is that religion perpetuates very bad ideas, and it's the only system of thought in which the process of

---

13 Emanuel Swedenborg (1688-1772) was a Swedish scientist, philosopher, Christian mystic, and theologian. Swedenborg had a prolific career as an inventor and scientist. At the age of 56 he entered a spiritual phase, in which he experienced dreams and visions. This culminated in a spiritual awakening, in which he felt he was appointed by the Lord to write a heavenly doctrine based on a reformed Christianity. For the remaining 28 years of his life, he wrote and published 18 theological works, of which the best known was *Heaven and Hell* (1758), and several unpublished theological works. He is just as well known for his scientific predictions, which were so far ahead of their time they were considered unworthy of mention at the time. From Wikipedia.

maintaining those bad ideas is considered a sacred act. The sacred act is disguised under the banner of faith. I have little doubt that religious faith will someday be seen as "*the most perverse misuses of intelligence we have ever devised*," to quote Sam Harris. Faith, as most people use it, is nothing more than a license to believe something that lacks evidence.

A lie told often enough becomes the truth.
Lenin (1870-1924)

Religion:

I would rather have a mind opened by wonder than one closed by belief.
Gerry Spence

A cause, principle, or system of beliefs held to with ardor and faith.[14]

Now we come to what I see as the real problem when we break it down to its fundamental elements, which is a system of beliefs held together with ardor and faith. We are watching ardor and faith in all of its "shock and awe" play out in all its glory and insanity in the world today and especially in the Middle East. I remember listening to Steven Weinberg at a conference when he said, *"Religion is like an old crazy aunt; she causes trouble, she lies, and we might miss her a little when she's gone, but it's time she goes."*[15] *He went on to say, "It's time we all wake up from the religious nightmare we've been living."* I know this sounds strong but believe me when I tell you this: Almost everyone I meet believes this to at least some degree but is afraid of saying anything for fear of persecution.

---

14 *Merriam-Webster's Collegiate Dictionary,* 2006.

15 Steven Weinberg (born May 3, 1933) is an American physicist. He was awarded the 1979 Nobel Prize in Physics (with colleagues Abdus Salam and Sheldon Glashow) for combining electromagnetism and the weak force into the electroweak force. Quote: "*Religion is an insult to human dignity. With or without it you would have good people doing good things and evil people doing evil things. But for good people to do evil things, that takes religion.*" *The First Three Minutes: A Modern View of the Origin of the Universe*, 1977.

The moment you accept your religion as the only way, it becomes an ideology and is a closed door. Eckhart Tolle

Some say that religion has burned like a candle flame through some dark times in our history (albeit they caused a lot of the dark times), but we are entering into the dawn of a new consciousness on our planet, a consciousness so bright that we no longer need the dim light of a candle burning in the night.[16] As harsh as you may think Steven Weinberg was, trust me, there are points of view out there much more inflammatory. Writers such as Richard Dawkins and Christopher Hitchens are quite abrasive with their comments about religions.[17] [18] I listen to them and others and I can see that everything they say is to a large degree true, but what they say is not going to persuade religious followers to change their divisive beliefs about themselves and "others." Christopher Hitchens writes quite candidly, "*Religion is violent, irrational, intolerant, allied to racism and tribalism and bigotry, invested in ignorance and hostile to free inquiry, contemptuous of women and coercive toward children.*" The *Bhagavad-Gita* says one of the most difficult things about fighting a war is to not become what you are fighting. One of my concerns and reasons for writing this book is to bring attention to this conflict between "reason" and "faith" before the situation gets to the point where the pendulum will swing too far in the direction of reason, leaving us in a cold and rational world in order to simply survive the madness of the world's belief systems and where they are taking us.

Morality:

Conformity to the rules of right conduct; moral or virtuous conduct.[19]

One of the religious believer's main points is that if we don't have a belief in a moral God who will punish us for our wrongs, then the whole world will run amok with atrocities of every kind. Scientists say that this is not the case. Every day more evidence shows that there may well be ways to

---

16 *Starseed Transmissions,* by Ken Carey, 1991.

17 *The God Delusion*, by Richard Dawkins, 2006.

18 *God Is Not Great: How Religion Poisons Everything,* by Christopher Hitchens, 2008.

19 From Dictionary.com.

deal rationally with human emotions and morals. Do I believe this? At this point there's not enough evidence to support this view one way or another so I must say, "I don't know," which is the crux of this book—having the courage to say "I don't know" and not to pretend that you do. What I do know is that to say that throughout history we have been morally guided by religious values would make most thinking people shudder. In fact, our Holy Books are in most cases far from being a moral guidebook. You will notice that I very seldom use the word "moral" because its definition is almost always based on the social environment of the time. When I was a kid growing up in West Virginia, it was considered your "moral right" to shoot and kill your wife and her lover if you found them in bed together, so we must use caution when we find ourselves even using this word. Besides that, even if there is a God giving us laws of how to be, this is not morality; this is a dictator dictating laws, which, from our point of view, have nothing to do with right or wrong and more to do with escaping punishment.

Socrates defended his morality this way: *"Do you think I'm an idiot? I have to live with these people."* He goes on to explain that he would not want to live among corrupt people and how other people act is dependent on how they are treated. So morality can and should be a matter of common sense and rationality. One problem we run into when we use religious beliefs as a factor in moral guidance is that if they are false beliefs (historically beliefs almost always prove themselves to be wrong or misleading), it's easy to use them to justify our actions based upon the belief in a particular outcome. The only safeguard is to ground our beliefs in nature and demand validation for them. They should be grounded in some type of "immutable truth," as Emmanuel Kant would say. It is easy to imagine the result if they are not—living in a world where morals and right and wrong end up taking sides and winning goes to the largest or most ruthless groups of people or nations. As a matter of fact, it is very easy to imagine that happening. It's been said that war never decides who's right or wrong, only who is left.

It doesn't take much when we look out in our world to see the effects of basing our morality on Holy Books and biblical principles. It will bring us to the edge of extinction if we keep heading in the direction we are heading; the terrifying part is that this is the exact plan of many of our Holy Books. Maybe it would be better to base our moral actions on a deeper understanding of the world around us, an understanding that we are part of a living organism called Earth, which in turn is part of a yet larger organism. We can exist only in groups; we are communal beings dependent on one another and, like it or not, we now live in a global village with other cultures, races, and religions.

To be totally fair when using nature as a guide for morality, we must take an objective look at it. For example, take chimpanzees. Male chimpanzees will slip off in the evening, go beyond their territory, and if they find another chimpanzee they will kill it in such a way as to make it suffer, even to the point of tearing its genitals off. Definitely there is something about nature that we must understand, and this is going to come about, not by reading a Holy Book, but through the natural sciences, such as biology. We are part of that nature but we have the ability to evolve and rise above it.

God:

We are all atheists about most of the Gods
that societies have ever believed in.
Some of us just go one God further.
Richard Dawkins

This is probably one of the most overused and misunderstood words of all time, and how could it not be? We are speaking of something completely beyond the realms of our understanding and comprehension, yet we seek to define it with our third-dimensional thinking and concepts. I hear people's constant misunderstanding of its use when it's spoken by people such as Albert Einstein and Steven Hawking. It's often used in such a way as to lend itself to the listeners' personal definition of the word. Scientists often use the word "God" as a general definition of something that could just as easily be called "infinite consciousness" or "the universe" or "nature" or "energy" or a long list of other ideas. Probably the biggest difference in definitions is whether or not we are talking about an impersonal God or a personal God who hears our thoughts and is a witness to our lives. In the final analysis, if we are being intellectually honest, there is no possible way to know who or what God is even if he or she does exist, let alone to know and be able to define with any accuracy the infinite being that either created this universe or "Is" this universe. Besides that, even if you do have a genuine encounter with an interdimensional being such as God, it will be of no real transformational value to another person. So you will probably notice that I will not be using this term much in this book. This book is not so much about what we believe in or even if it's true or not, but how and why we believe at all when there are so many avenues of knowing with today's advances in technology, science, and the many, many archaeological finds in the past few years, from the Dead Sea Scrolls and Nag Hammadi libraries to the Sumerian clay tablets, almost all of which contradict our traditional beliefs, especially the ones of a vengeful, capricious, and jealous God that we can barter with, as in the case of Sodom and Gomorrah.[20]

Another thing to keep in mind is that what we call God has changed many times through the centuries. We would be challenged to find people who still believe in Zeus or Poseidon, so in that sense we are all atheists in regard to those particular Gods. Zoroastrianism used to be one of the world's largest religions and I have never met or heard of one person who follows that religion and probably never will.

One more thing to think about as far as defining the creator of our universe is to first recognize our physical placement in the scheme of things. We live on this tiny speck of dust floating in the night sky, orbiting around a sun that is one of trillions of suns in billions of galaxies and everything we know, all of our experience, our history, everything comes from this speck of dust. It's only human arrogance that would allow for even the possibility that we know anything at all in the way of an ultimate truth about our existence. It's no wonder that many of us have found comfort in pretending to know something about the creator of the universe and made him or her our personal God who watches over us.

---

20 In Genesis 18, God informs Abraham that he plans to destroy the city of Sodom because of its gross immorality. Abraham pleads with God not to destroy Sodom, and God agrees that he would not destroy the city if there were 50 righteous people in it, then 45, then 30, then 20, or even 10 righteous people. The Lord's two angels found only one righteous person living in Sodom, Abraham's nephew Lot. Consequently, God destroyed the city. What may be of interest here is that a few classical Jewish texts do not specifically indicate that God destroyed Sodom and Gomorrah because the inhabitants were homosexual, or sexually deviant from what was recorded as God's law of natural order, but rather, they were destroyed because the inhabitants were generally morally depraved and uncompromisingly greedy. Rabbinic writings affirm that the primary crimes of the Sodomites were terrible and repeated economic crimes, both against each other and outsiders.

Spiritual Evolution:

The further the spiritual evolution of mankind advances, the more certain it seems to me that the path to genuine religiosity does not lie through the fear of life, and the fear of death, and blind faith, but through striving after rational knowledge.

Albert Einstein (1879-1955)

This term, once fully understood, could in itself change the way you see the world and your relationship to it. Spiritual evolution could be understood as the philosophical, theological, esoteric, or spiritual idea that nature and human beings and/or human culture evolve along a predetermined cosmological pattern or ascent, or in accordance with certain predetermined potentials. This points to a predeterminism in the evolutionary concept; however, spiritual evolution has the creative human impulse as a variable, called epigenesis. The definition for spiritual evolution as it is used in this book is simply this: The underlying matrix or fabric of life is infused with a consciousness that has, from the beginning, had the impulse to evolve into higher states of being, allowing even greater consciousness to come through. Epigenesis is important to understand in this respect because without this impulse in evolution it's very doubtful that human beings would have evolved to our present state of being self-aware.

Pierre Teilhard de Chardin, the Jesuit paleontologist, refers to this as the Omega Point and the coming together of mankind in what he refers to as a biosphere. In other words, we may be at a point in the spiritual evolution of humans when we connect into a higher form of being, a planetary being, which of course is speculation. What is not speculation is the fact that we have evolved to this point, which is where we are starting to become active and conscious participants in our own evolution.

CHAPTER 3

A Few Words to Believers and Nonbelievers

Since I review books and interview best-selling authors for my magazine I know there's an abundance of positive information out there to help us change our lives by changing our thoughts. The idea is built around "what we focus on is what we create in our lives." This is all very true but I have noticed through the years, as I'm sure you have, that all the positive affirmations and good intentions in the world simply don't change some people's lives. (For your convenience I have included a list in the bibliography of wonderful authors who offer this type of information.)

The reason for this, as it turns out, is quite understandable in terms of our conscious awareness and our unconscious "tapes" that play in the background. These tapes are almost always tied to our beliefs of ourselves and the world we live in. If we don't put light on those limiting unconscious beliefs, our conscious minds don't stand a chance, because our conscious minds occupy around 5 percent of our mental power to make decisions and our unconscious minds occupy the other 95 percent and are far more powerful. The unconscious mind is the part of our minds that most people say we don't use at all, which may be true, but it uses us.

It's beyond question that the whole of our worldview and the destiny of our civilization rests in our beliefs. Our personal beliefs about our existence, who God is, and how we got here motivate most of our decisions in the world today. So my question is this: *"What if our beliefs are wrong?"* Doesn't this possibility warrant at least some attention? What if the writers of our Holy Books weren't inspired by God but inspired by nationalism and racial ambitions and agendas?

What we must do is go past the separate beliefs that we hold as individuals and nations; we must look at the cognitive process of "believing" and consider the possibility that we are evolving beyond the need to *"pretend we know something that, in reality, we have no way of knowing."* Of course what makes believing even more dangerous is that we eventually forget we are pretending and assume we "know." (*This paragraph is worth reading again.*)

Finding the Common Denominator

What if your child were hit by a drunken driver? Would you be concerned with what kind of car it was? No. You would want to know who was driving and what mental condition he or she was in at the time. Was it an unforeseeable accident or was the driver drunk? It's in this light that we must start viewing what's going on in the world today. It doesn't really matter which religion or political regime is taking today's world stage of absurd and insane actions; what matters is the common denominator that has fueled these conflicts for millennia—which is their beliefs.[21] At the root of most of the world's insanity you will find people who are drunk with a belief in an invisible God whom they read or heard about from a book that was supposedly written or inspired by the same invisible God. Of course this becomes even more dangerous when the same invisible God gives them the divine command to "spread the word" by conversion or else, and it's almost always "or else" because no one wants to give up his or her beliefs.

Intervention for Believers

In a sense you might look at this book as intervention for believers in the same way we have intervention for drug addicts and alcoholics. And if you are a believer, please do not let me lose you here. Stay with me because it will be worth it. Most addicts have a difficult time seeing and admitting

---

[21] I will be using the word "absurd" frequently throughout this book so I want to clarify it. The origin of the word dates to 1550-1560 to *absurdus* and meant "out of tune." If you look it up in today's dictionary it means "utterly or obviously senseless, illogical, or untrue; contrary to all reason or common sense; laughably foolish or false: an absurd explanation. The quality or condition of existing in a meaningless and irrational world." So when I use this word I'm using it literally and not in a derogatory manner. From http://dictionary.reference.com.

they have a problem, which is a well-known phase called denial. What it usually takes is a life-threatening traumatic event and admitting what their actions have caused. Sometimes it is an accident, loss of their families, or even death. So that is what we are going to look at—the consequences of some of our beliefs. In this way we may be able to shine the light of awareness inward and see the beliefs that caused these consequences, but it is important to first see what those consequences are. This may not be easy, but as I mentioned before, it will be worth it.

Also, the purpose of this book is not to simply give you something else to believe in; rather, it's to shine light on the many absurd and limiting beliefs that have been handed down to us by the world we live in, our parents, religions, governments, teachers, philosophies, ideologies, and the list goes on. But without relinquishing our attachment to these deep-seated beliefs we will never move forward and more than likely will move backward. But this will never happen until we start looking at the mental process of our need to believe at all.

When talking about absurd beliefs it is difficult not to come across a little strong at times, and I know that at times you may hear an undertone of something that seems like anger, but it's not. It's a deep feeling of urgency. If you observed a child running straight into a flow of traffic and the only thing you could do was yell, it would be difficult to yell for the child to stop without sounding angry. It's a matter of urgency. Be assured that I am making every effort to come from a place of compassion with the full understanding that we are all one and that if I were in your shoes with your life experiences, education, and indoctrination into a specific belief, I would probably be doing the same things you are and believing the same things you do.

CHAPTER 4

The Mental Disconnect

Another intention of this book is to show the necessary and inevitable "mental disconnect" involved in supporting beliefs that have no empirical evidence and indeed, go against all known science and experience, and even common sense most of the time. For instance, neither you nor I have ever witnessed a virgin birth as taken literally in the Bible. No one has ever seen a winged horse such as the one Muhammad is supposed to return on. And I know that I have never heard a booming voice declaring it was that of God coming from a burning bush; yet we base the validity of our beliefs on such notions being true and hundreds of millions of people organize their entire lives around these beliefs.

Obviously this list of the beliefs we hold to be true and sacred and that we are not allowed to question even within our own religions could go on almost indefinitely. Of course we always see the absurdity in other religious and political beliefs. But let's put aside whether or not they are true or not and focus on the cognitive process of "believing" for a moment. What is this seemingly insatiable desire for pretending we know something that we have no way of knowing? Actually the only thing we do know is that we have no way of knowing; that's why we say it's based on faith. This could just as easily be the rationale and reasoning that a madman would use to validate his delusions.

I ask you to at least entertain the possibility that this is a leftover cognitive function of what may someday be viewed as a primitive and unevolved mind. I say this because this idea of believing dates to our earliest times in antiquity, when we were trying to make sense of a world of which we knew nothing; the only thing we have to counteract this is our strict adherence to the scientific process of inquiry, validation, and education. Even now we equate how primitive a society is to the degree its members have held on to superstitions.

This isn't to say that intelligent people do not believe in religious myths as a literal truth; numerous religious leaders are very intelligent and it's to them that this book could be very helpful if they read it. If you are one of their "flock," then by definition you have already handed over your right to think for yourself and are either being led by a charismatic leader or a sacred scripture. If you happen to be a follower, you might find it interesting that many religious leaders are waking up to what we will be addressing and I have interviewed some of them for this book. So if you don't want to be "left behind," keep reading.[22]

If the truth were known, I probably would not be writing this book except for the fact that I don't see anyone addressing some of these issues and I see us heading for a collision with some self-fulfilling apocalyptic prophecies if we continue unabated on this course. I would much rather write something that would put me in the mainstream of the movement I see happening on this planet and feel very much a part of, one that could be identified as a spiritual movement toward peace, love, and oneness. However, the days of hiding with our heads in the sand are over, and time is not on our side.

There are two types of evolution: collective evolution and individual, conscious evolution. "Evolution" implies unconscious, collective progress, so it would be better to use the word "revolution" in talking about man. With man, revolution becomes possible. Revolution, as I am using the word here, means a conscious, individual effort towards evolution. It is bringing individual responsibility to a peak. Only you are responsible for your own evolution.

Osho (1931-1990)

---

22 The Left Behind series of books talk about end times, rapture, and apocalyptic wars in which the nonbelievers are "left behind" and Christians are whisked away to heaven. Tim LaHaye, the author, isn't merely a fringe figure. The former cochairman of Jack Kemp's presidential campaign, LaHaye was a member of the original board of directors of the Moral Majority and an organizer of the Council for National Policy, which ABC News has called "the most powerful conservative organization in America you've *never* heard of" and whose membership has included John Ashcroft, Tommy Thompson, and Oliver North. George W. Bush is still refusing to release a tape of a speech he gave to the group in 1999. These are powerful people who seem comfortable with a story about the world coming to an end . . . our world.

Whether or not we are approaching a point in our evolution called planetary awakening, or even massive individual awakenings, is not clear; what is clear is that we need to change the direction some of us are headed. One thing I am concerned with is some of the details, such as the fact that literally hundreds of millions of people on the planet believe that the sooner we devastate and destroy our environment the sooner a savior, theirs by odd coincidence, will come back and whisk them all off to paradise.

In olden days, when our technology consisted of a sharp spear and implements of torture, the only thing a rational person needed to do was keep quiet and stay out of the way of soldiers with large crosses on their shields. But things have changed now that we live in the atomic age; you cannot duck or jump out of the way of an atomic bomb or a 747 aircraft plowing into a building. We no longer have the luxury or option of waiting 1,000 years as we did during the Dark Ages. We must, all of us, speak out even if it firmly puts us on the fringe of a small percentage of the population of this planet. We must also keep in mind that in today's world it takes only one person to kill a million people, and people are waiting in line for the opportunity to do so. This is a first on our planet, so it requires the urgent attention of people who are not guided, or misguided, by an invisible being who seems to speak only to them.

Any rational person knows that we may be sitting on the only inhabitable planet in our known universe. If we continue on this course we will most assuredly destroy any chance we have of living on it. This affects all of us and our children. I have been told by well-meaning spiritual people and friends that looking at these problems is being negative. I will agree that it's negative, but it's not the "looking" that's negative; it's what we are looking at that's negative. If I shine a light into the darkness and it lights up an ugly mess, we can't blame the light. The only way to transform negativity is to put conscious light on it. Nothing generates conscious light faster than self-reflection and focusing that awareness on negativity without any judgment. Negativity in itself has no power to transform or to see itself, which can be evidenced in any number of scenarios in our world today. The only way to change the things that we will be looking at is by putting conscious light on them. This is how the insanity has gone on for so long—we have been intimidated into submission and afraid to question it. Those days are gone as far as I'm concerned and I invite you to join me. We will go much deeper into the mental disconnect a little later in this book.

CHAPTER 5

My Invitation to Religious Believers

It is my sincere wish that you read this book with all the intellectual honesty you can bring forth and begin to think for yourself. We can accomplish this only by looking honestly at some of the things we believe and asking the tough questions that, for some reason, most believers don't ask, such as: *"Why do I believe what I believe and what gives me the authority and right to impose those beliefs on others even if it risks the very planet that we all inhabit?"* This is an invitation, and maybe a challenge, to you to answer that question honestly without escaping into age-old mental gymnastics to avoid answering.

Using faith alone as a reason to believe something absurd can easily be identified and disqualified by any believer when looking at another belief system, but we always fail to use the same guidelines and restrictions on our own beliefs. We will be looking at some very uncomfortable ideas and beliefs that we have held on to for far too long. Observe the fact that our three dominant world religions had their births in the Iron Age, when the people who founded them had no scientific understanding of the universe or even the world in which they lived. It's no wonder that our world religions cause so much suffering to others and to ourselves when they are based on such a limited understanding of rudimentary knowledge that most high school students now have.

Compassion Is the Basis of All Morality

A question that we often ask ourselves is, "How is it that man can cause so much suffering to another and at the same time, turn around and give his life to save someone else that he doesn't even know?"
Arthur Schopenhauer (1788-1860)

According to Schopenhauer there is a breakthrough of realization that "you" and the "other" are *One* and that the separateness that we normally perceive is an illusion and is an effect of the sensibilities of time and space.[23] Our true reality is Oneness and when we perceived that, in times of emergency for instance, this realization overrides our normal biological need for self-preservation. This obviously means there is something deeper than the physical part of our being.

A good example of this was a story I heard on the *Oprah Winfrey Show* about a small boy who was going to commit suicide by jumping off a bridge. A police officer showed up and was able to grab hold of the boy just in time but he was also being pulled over. If it weren't for the second officer's showing up, the first police officer and the boy would have both gone over. Later, when the first officer on the scene was asked, "Why didn't you let go?" he answered, "I couldn't." Something in him simply couldn't let go of the boy. This can be understood only through realization. We need to bring this realization into awareness on a global level if we expect to survive as a human species.

This elevated and selfless consciousness that I am referring to is important to keep in mind as we look at the many crazy and absurd things the "other" people believe in. We must, not even for a moment, let ourselves fall into anger or judgment, but rather always come from a place of compassion. Then and only then will we know what the police officer meant when he said he simply couldn't let go.

I remember listening to Bill Moyers in an interview with Joseph Campbell asking, *"How do you love your enemy?"* Joseph Campbell said, *"You*

---

[23] A key aspect of Schopenhauer's thought is the investigation of what makes man less than reasonable. This force he calls *Wille zum Leben* or Will (literally will-to-life), by which he means the forces driving man to remain alive and to reproduce, a drive intertwined with desire. This Will is the inner content and the driving force of the world. For Schopenhauer, Will had ontological primacy over the intellect; in other words, desire is understood to be before thought, and, in a parallel sense, Will is said to be before being. Schopenhauer thought this was similar to notions of *purushartha,* or goals of life, in Vedanta Hinduism. Schopenhauer's moral theory proposed that of three primary moral incentives, compassion was the genuine motivator to moral expression. He ruled the other two, malice and egoism, corrupt as moral incentives. The identification of compassion as the true moral incentive was a central aspect of Schopenhauer's mission. From Wikipedia.

love them by taking care of yourself like it says in the Bible: . . . How can you say to your brother, "Brother, let me remove the speck that is in your eye," when you yourself do not see the plank that is in your own eye? Hypocrite! First remove the plank from your own eye, and then you will see clearly to remove the speck that is in your brother's eye." (Luke 6:42)

We have, in fact, two kinds of morality side by side: one which we preach but do not practice, and another which we practice but seldom preach.
Bertrand Russell (1872-1970)
"Eastern and Western Ideals of Happiness," *Sceptical Essays* (1928)

So as we continue on our journey let's remember to consider all human beings as sacred and spiritual. I might add that I have just alienated myself from most of the atheists of the world but I see them as the necessary knee-jerk response to religious fundamentalism. It's just as senseless to say that there are no mystical experiences and no God simply because you "don't" believe in them, which is a belief in itself. Somewhere in the middle of all this is a path that will lead us to a deep *experiential* understanding of our reality. But we must let go of our beliefs and even the need to have a belief when that same belief stops us from exploring and researching to find answers that allow us to "know" the truth, not simply "believe" we know the truth. We must learn to live in the mystery called existence and be content with figuring out this puzzle, one piece at a time, and stop pretending we know things we do not.

So if you are a religious believer I ask you to trust the fact that I'm not out to talk down to you or at you, but *with* you, human being to human being, living on this isolated planet called Earth. We are all together on this journey, wanting to live out our lives in peace.

CHAPTER 6

Paradigm Shifts

So far our history has experienced several epic moments, the realization of which was so profound that they changed human consciousness on a global level.

- One was in the 6th century BC, when Pythagoras said, *"The earth is spherical, not flat."* Of course, later on, this information was suppressed for many centuries by the Church.
- Another was in the 16th century AD, when Copernicus said, *"The earth revolves around the sun."*
- Then we had another in the 20th century, when Einstein came out with his formula E=mc2, showing that energy and mass are two different expressions of the same thing.

Now, in the 21st century, we are living through the next quantum leap in our ever-expanding consciousness as we realize that *"physical reality is a manifestation of thought."* Volumes of books and scientific papers validate this view so it's not just some wild idea coming out of the mind of some theoretical scientist.[24] I might also point out that every one of the previous ideas mentioned was violently opposed and suppressed by the same religious organizations that still exist today, and large parts of the population were

---

[24] The famous "double slit experiment" is one of many experiments that clearly illustrate how our consciousness has a direct effect on matter. In thousands of experiments it has been shown that light is a wave until it is observed, at which point it turns into a particle. The implications of this on both a philosophical and scientific level is that we create reality and interact with it on a conscious level. A good website to observe this is www.doubleslitexperiment.com.

kept in the dark, persecuted, executed, tortured, and thrown in prisons over some of those ideas.

Philosophers and psychologists have had great difficulty understanding how an inert, dead matter can lead to living consciousness. But that's not how it happens. In our quantum view of the universe, consciousness is ubiquitous. Intelligence is everywhere. And the deeper you go beneath the surface, the more intelligence, the more dynamism, the more awareness, until at the foundation of the universe there is a field of pure, abstract universal existence—Universal consciousness— The unified field.

John Hagelin[25]

Awakening to Self-Awareness

If we go back even further in our history, we can see that there was an awakening of consciousness in which we became self-aware, as recorded in the Sumerian Tablets and later in Genesis. This was when we approached that inner awareness of "I am" and stood in wonder of our existence deep in this mystery called life. It may be one of the first human insights into our nature that is described in ancient scriptures, which told through analogies and stories how we looked down and saw our nakedness in the Garden of

---

[25] John Hagelin, PhD, is a world-renowned quantum physicist, educator, public policy expert, and leading proponent of peace. Dr. Hagelin received his AB summa cum laude from Dartmouth College and his MA and PhD from Harvard University, and he conducted pioneering research at CERN (the European Center for Particle Physics) and SLAC (the Stanford Linear Accelerator Center). His scientific contributions in the fields of electroweak unification, grand unification, supersymmetry and cosmology include some of the most cited references in the physical sciences. He is also responsible for the development of a highly successful Grand Unified Field Theory based on the Superstring. Dr. Hagelin is therefore at the pinnacle of achievement among the elite cadre of physicists who have fulfilled Einstein's dream of a "theory of everything" through their mathematical formulation of the Unified Field—the most advanced scientific knowledge of our time. He also appeared in the movie, *What the Bleep Do We Know?*(2004). For more information visit his website at www.hagelin.org.

Eden and realized our existence, "I am."[26] I often ponder the wonder of those first individuals who looked into the eyes of some of their fellow human beings and noticed that there were some who were aware and some who were not. In the chapter on 2012 we will cover some possibilities of what brought this awareness about; they may surprise you.

Ever since those first impressions of our existence we have been identifying "thinking" with "I am," and this has proved to be very misleading for anyone seeking his or her deeper essence. "*I think, therefore I am*" is but one of the many outdated paradigms. In fact, thinking is the very thing that needs to stop before we can peer behind the veil of illusion with pure consciousness into the inner secrets of our true existence.

Since that time we have dealt with this egocentric awareness in the most peculiar and bizarre ways, which, all things considered, was inevitable. But now we have come to another summit of human conscious transformation where we are looking "in," and now we see our inner spiritual connectedness with existence, and we are still trying to cover ourselves. Except now, instead of fig leaves as written in Genesis, we cover ourselves with layers of ego personalities based on our possessions, our race, our country, and most dangerous of all, our religious beliefs, all of which blind us to the hidden meanings behind sacred texts. Some of the earliest observers of this emerging consciousness were Lao Tzu, Buddha, and Jesus. But I have yet to run into one Christian who understands the deeper esoteric meaning of "being one in the body of Christ" or who has even attempted to love his enemy. That's also true of the hidden meanings behind many of the parables such as the resurrection, crucifixion, and virgin birth. We have lost the meaning because we took the message literally.

The virgin birth is symbolic of the birth of spiritual man out of the animal man through the compassion of the heart.
Joseph Campbell (1904-1987)

---

[26] Genesis 3:5: For God doth know that in the day ye eat thereof, then your eyes shall be opened, and ye shall be as gods, knowing good and evil. 6: And when the woman saw that the tree was good for food, and that it was pleasant to the eyes, and a tree to be desired to make one wise, she took of the fruit thereof, and did eat, and gave also unto her husband with her; and he did eat. 7: And the eyes of them both were opened, and they knew that they were naked; and they sewed fig leaves together, and made themselves aprons.

It's been said that we should always speak from our own experience. This is an accounting of the information and personal experiences that have led to a transformation in how I see something in this world called "beliefs and the process of believing." Most of the beliefs we have about our existence are rooted in the experiences of others who lived, or may have lived, thousands of years ago. Expecting to get something out of the experiences of others is much like expecting to be nourished while looking at photos of other people eating; it simply doesn't translate into anything meaningful or nourishing and in the end leaves you hungry. At first, my own transformation was almost imperceptible and seemed irrelevant, except for a slight sensation of inner freedom, but it has had a deep impact on how I see the world, both inner and outer.

Given our ancestors' lack of scientific knowledge, imagine for a moment what it must have been like for our distant relatives as they looked into the night sky and watched the stars and wondered what they were. Some of them who watched the skies methodically through the years noticed that some of those points of light "wandered" so they must be Gods; the ancient Greek meaning of the word "planet" is "wanderer." They (the stars) became some of our first Gods in a long list of Gods that have long since gone to the religious graveyards we call "mythology." In reference to those long-lost Gods we are all atheists now. This isn't to denigrate our ancestors; it was only natural to make an attempt to explain the world they lived in—a world they had very little understanding of or control of and oftentimes became the victims of by a seemingly vengeful, jealous, and capricious God. And still there are those who think nature is God's handiwork to take revenge on the wicked. I remember Jerry Falwell saying the reason hurricane Katrina hit New Orleans was to punish the homosexuals.[27]

Science without religion is lame, religion without science is blind.
Albert Einstein (1879-1955)

But with the rise of technology and time to measure and investigate, science undermined the beliefs of the day, that those points of light were Gods

[27] Jerry Falwell (1933-2007) was an evangelical pastor and a conservative commentator well known for his political connections, his fierce opposition to the women's movement, and his controversial statements blaming gays, feminists, pagans, and abortionists for the September 11 attacks.

and the sons and daughters of Gods. Today you would need to search the world over to find more than a handful of people who still believe the stars are Gods because, thanks to science, we now have a better understanding of what they are. They are nuclear fires of hydrogen and helium created by enormous forces of gravity pulling everything toward their center until nuclear fusion happens, releasing a photon speeding through space and time at the speed of light. Nevertheless, many people killed and died for those long-lost beliefs in their Gods, thinking all the time they were somehow appeasing their loving God.

One of the greatest pains to human nature is the pain of a new idea.
Walter Bagehot (1826-1877)

Does it diminish our mystery when rational men of science figure these things out? Of course not, and most any educated human being would agree that their discoveries always take us to new heights of understanding and to even deeper mysteries by releasing us from outdated and limiting beliefs that held us back from further inquiry. Did some of the people in those times have difficulty handling this transition? To be sure, it must have been horrifying for some to listen to the heretics of their day who no longer believed those lights in the sky were Gods. They were surely going to suffer the wrath of Zeus, Osiris, or Poseidon for having such evil thoughts. We've experienced many such transitions in history, usually brought about after many years of questioning and, of course, the fact that Zeus never showed up and hit the heretics of the day with a bolt of lightning.

Religion is a defense against religious experience; concepts and doctrines keep a person from having a transcendent experience.
Joseph Campbell (1904-1987)

There was a time when it was considered immoral, and therefore illegal, to cut up a corpse even for medical inquiry—the reason being that this was the domain of God and by looking into it and studying it we would take away from the mystery somehow. Of course anyone who has ever studied physiology or biology knows all too well that there's plenty of mystery left, and in fact, with every answer we seem to find ourselves even deeper in the mystery. But like most beliefs, they change and the last remnants die one grave at a time, but the main transition can happen within a generation or two. And, more important to you, on an individual level it can happen in less than a heartbeat.

It is worth recalling that it took brave pioneers many years to overcome the powerful taboo against the dissection of human cadavers during the early years of modern medicine. And we should note that, notwithstanding the outrage and repulsion with which the idea of dissection was then conceived, overcoming that tradition has not led to the feared collapse of morality and decency. We live in an era in which human corpses are still treated with due respect—indeed, with rather more respect and decorum than they were treated with at the time dissection was still disreputable. And which of us would choose to forgo the benefits of medicine made possible by the invasive meddling science?

Daniel Dennett

Breaking the Spell (2007)

Now we find ourselves confronted with a similar situation with more of our cherished beliefs being threatened. The heretics who we call scientists and cutting-edge spiritual thinkers are stealing our notions of what and who God is and indeed, who and what we are as beings in this universe we find ourselves in. Again, I must ask the question: *"Could it be that we are finally evolving beyond our need to pretend we know truths when in fact we have no idea if it's true or not?"*

I also want to remind you that my purpose in writing this book is not to simply attack religious beliefs. I'm talking about *all beliefs* that we cling to that *distort, limit,* and *alter* the way we process information about ourselves and the world we live in. This includes beliefs such as you're not worthy of being loved because when you were a child you were abandoned by a parent or parents, or the belief that sex is the original sin and therefore viewed with disgust. We have experienced more than 70 generations of sexual suppression in the name of a loving God who supposedly gave us the urge to make love in the first place. This is also about our national belief that our country could never be part of something wrong and to question that means you are not a patriot. We have many deep-seated beliefs that are not connected with reality and they live on because we no longer question them and continue to believe them in the face of hard evidence to the contrary. What adds even more confusion in the realm of religious beliefs is that religious followers have not read their Holy Books and so they're not even familiar with unsettling details, such as *they are commanded to kill their children if they are disobedient or drink a little too much.*[28] And of course we must not leave out researchers and scientists who routinely discount new discoveries because they don't fit the prevailing theory of the day.

One of the attributes of a hero in any myth
is that he is ready for the adventure.
Joseph Campbell (1904-1987)

During our adventure together we will look at just how we set up neural pathways, or what some scientists call imprints, that lay the groundwork for believing and that are deeply ingrained in the structure of our brains and affect our biology on a molecular level. This starts at a very early age, when we tell our children strange and absurd beliefs that completely disconnect a child from reality. I'm not referring to the rich and imaginative mind of a child; I'm referring to the stories we are told as children by the power figures in our lives, stories that are completely absurd in regard to reality and that are told as truths to believe in and not to be questioned. We constantly indoctrinate our children into our beliefs, which is how beliefs continue generation after generation. A good documentary to watch is *Jesus Camp* (2006), in which fundamentalist Christians are indoctrinating very young children to be "Christian Soldiers" to fight the "good war" against nonbelievers. It's horrific to watch and it's going on right here in the United Sates of America, not in Afghanistan, Iraq, or Israel. What's ironic is the people in the documentary thought they were being filmed for doing something profoundly good and important. We would never call our 5-year-old sons or daughters Republicans or Democrats or independents. Why? Because we think they are far too young to know where they stand on such issues, yet we are fully confident in their ability to know the deepest

[28] Deuteronomy 21:18-21: If a man have a stubborn and rebellious son, which will not obey the voice of his father, or the voice of his mother, and that, when they have chastened him, will not hearken unto them: Then shall his father and his mother lay hold on him, and bring him out unto the elders of his city, and unto the gate of his place; And they shall say unto the elders of his city, This our son is stubborn and rebellious, he will not obey our voice; he is a glutton, and a drunkard. And all the men of his city shall stone him with stones, that he die. (King James version)

and most profound cosmological truths of the universe, truths that have challenged and eluded the greatest thinkers and philosophers throughout time, by getting them to claim they are Christian, Jew, or Muslim.

We will also look at some of the evidence that believing is deeply ingrained in us through evolution and exactly what are some of the evolutionary origins of our need to believe in a God. In my conversation with Andrew Newberg you will hear evidence that points to the possibility that we are hardwired to believe in a God.[29] Of course if you happen to be a "believer" you have long since given up the need to inquire into the origins and legitimacy of your beliefs. If that describes you, I might remind you that history has shown almost every belief we've ever had has turned out to be either completely wrong, or at best, misleading. Our need to believe has led to a long succession of new and unending beliefs.

With most men, unbelief in one thing springs from blind belief in another.
George Christoph Lichtenberg (1742-1799)

So before you turn the page what you need to ask yourself is: *"Do I have the courage it takes to be intellectually honest enough to read a book that may take some of my most cherished beliefs?"*

---

[29] Andrew Newberg, MD, is an assistant professor in the Department of Radiology at the Hospital of the University of Pennsylvania with secondary appointments in the Department of Psychiatry and the Department of Religious Studies. Dr. Newberg's research now largely focuses on how brain function is associated with various mental states, in particular, the relationship between brain function and mystical or religious experiences. The results and implications of this research are delineated in Dr. Newberg's best-selling book *Why God Won't Go Away: Brain Science and the Biology of Belief* (Ballantine/Random House, 2002); www.andrewnewberg.com.

CHAPTER 7

Dhara Lemos, Linda Francis, Gary Zukav, and Rahasya

Looking for Our Authentic Selves
A Conversation with Gary Zukav

My wife and I had the pleasure of meeting Gary Zukav and his wife/partner Linda Francis in Seattle, Washington, during a Spiritual Partnership workshop. Gary is an excellent example of how a person can change his life by changing his thoughts and beliefs. My first experience with Gary Zukav was many years ago through his book *The Dancing Wu Li Masters,* which is a comparison of Eastern thought and philosophy with Western science. But it's been some of his latest work in the area of sacred partnerships that has captured the attention of the evolving collective social consciousness. In a sense, you, the reader, and I have a sacred partnership because we are sharing our time together with reverence and a respect for truth.

I was especially interested in what Gary had to say about the beginnings of religious and social organizations since many of them today, particularly religious organizations, had their beginnings thousands of years ago. This was a time in our history that's difficult for most of us to comprehend unless you happen to live in an area that is still controlled by these social and religious beliefs, as in a theocratic state. What follows is my conversation with Gary on this and related issues.

Rahasya: When you speak of a transformation of consciousness on a global level, which is actually our collective consciousness, I wonder if there have been times in our evolutionary past when this has also taken place on a different level. For instance, looking back in our history it seems that human beings went through a transformation of consciousness when we reached a level of awareness of our own existence and basically said with conviction, "I am." This could be the metaphor of Adam and Eve in the Garden of Eden when they looked down and saw their nakedness and covered themselves. I think what we covered ourselves with was layers of personalities from which we are still emerging. Now we seem to be looking inward and seeing our inner spiritual existence. We are actually becoming, seeing, and feeling, and sensing that spiritual being, and as we sense it, we are also sensing the unity and the oneness of it. And it's putting us in touch not only with other individuals but it's putting us in touch with the planetary being that we are part of. Do you see it this way?

Gary Zukav: All those things are happening, and at the same time, the creation of authentic power is something that you have to do for yourself. I have to do it for myself, to align my personality with my soul to create harmony and cooperation and sharing and reverence for life in my life, which requires me to come into awareness of all of the parts of personality that are based in fear, and by that I mean the parts that are angry, jealous, resentful, anxious, frightened, that feel superior, or that feel inferior, the need to please, that are compulsive, that are addicted, and it's necessary not only to encounter and become familiar with all of these parts of your personality but to challenge them and to feel them. And that is spiritual growth, which is the creation of authentic power. And that does not happen automatically. It takes a lot of work, it takes emotional awareness, it takes responsible choice, which means choosing your intentions consciously and wisely. It takes consulting your intuition. It takes trusting in the universe. And all of these things, all of these skills, are the skill of creating authentic power. And the reason that I have become so interested in spiritual partnership is the same reason that hundreds of millions of other individuals are now becoming interested in it, whether or not they even know that there's a name for it, because as we become a multisensory species, our evolution is no longer based on physical survival but spiritual development.

Rahasya: I would imagine that most evolutionary planets reach this point where their survival is dependent on spiritual evolution, especially during their technological adolescence, which is where we find ourselves. We are sensing that we are connected energetically and spiritually with every other living being.

Gary Zukav: And so a new archetype of relationship has come into being that serves a multisensory species that is evolving through spiritual growth. And that new archetype is spiritual partnership. A spiritual partnership is a partnership between each of us for the purpose of spiritual growth. Nothing like this archetype has existed before because nothing like this was required before in the human experience, but it exists now because we have crossed an evolutionary threshold, and we're becoming aware of ourselves; we are becoming aware of larger aspects of ourselves, mainly our soul. And as individuals begin to understand and long to align their personalities with their souls, at the same time they start to long for relationships with substance and depth, in which they talk about more than what they read in some books or how the kids are doing, or what politics is, or how their work is going. Those things are all important, but in these relationships of substance and depth they also discuss the emotions and the challenges and the aspirations that are most important to them, and they help each other to grow spiritually, to create authentic power by becoming more and more open and involved themselves. That means more emotionally aware, more self-responsible, cultivating their intuition. And that is the reason why I love spiritual partnerships. And spiritual partnerships can be created in a biological family, they can be created among friends, they can be created in the workplace, they can be created anywhere that two or more individuals are committed to their own spiritual evolution and are striving to relate to each other as equals.

Rahasya: This is a touchy question for some people when it gets into this area because historically we have given religious believers a special dispensation from being questioned about their beliefs. But it seems to me that a large proportion of today's world problems are centered around people's religious beliefs, which is no surprise when you look at history. So I think we need to start holding people accountable for their beliefs. But something else I've noticed is there seems to be a limitation in understanding for what many of us see as an obvious fact. And then the first thing you know you are building a $27 million dollar museum that teaches children the world is 6,000 years old. I heard the Dalai Lama say once that if it ever comes down to scientific evidence or Buddhist beliefs, science will win. What are your thoughts on belief systems that are no longer open to inquiry or questioning?

Gary Zukav: They are based in fear. The extent to which frightened parts of their personality control them and it is to this same extent to which they are rigid and inflexible. Therefore, when you encounter a person who's a fundamentalist of any kind, Christian, Muslim, new age, whatever, who is uncomfortable with people who do not believe as he does, or talk, or think or behave as he does, or act as he does, or even dress or look as he does, then you are encountering a frightened person. This kind of person is so rigid and so inflexible that they are not open to any other perspective or point of view; then you are speaking to a very, very frightened person. All fundamentalists are very frightened people. They are rigid and they are inflexible; this is the essence of fundamentalism. A Muslim fundamentalist and a Christian fundamentalist have the same fear. They are indeed brothers and sisters in fear. They are also brothers and sisters in love but they don't realize that yet.

Then we come to the topic of organized religious belief systems. They are each a cultural creation, they are each expressions of universal truths, but nevertheless through cultural filters. That is, religions are cultural events and cultures are spirit-based collectives that separate us from them. In cultures, a spirit that's corrupted will separate "us" from "them." They [religions] provide refuge for those who are like ourselves and allow us to defend ourselves against those who are not like us. That's the origin of culture. And religions are culturally created and the genesis of religion is in culture. And since cultures are fear-based collectives then so are religions. And you can see for yourself that this is so because religions, if you look at the history of religion and look at what they are doing in the world now, notwithstanding the people with high ideals in religions, we see that religions are divisive, they divide us, they cannot unite us. Each religious individual, whether that individual is Christian, Muslim, Hindu, or Buddhist, thinks in his heart that he has the inside track, that he has the real truth, that he has the ultimate knowledge and no one else does. That makes it difficult to have tolerance, so there is no future in terms of the universal human that is longing to emerge now as a multisensory humanity by way of religion. You cannot create a true universal humanity by imposing one culture on another; that's the pursuit of external power, by the way, the ability to manipulate and control. So you can never create a true universal humanity by attempting to impose one belief system, or a religion, on others.

Rahasya: This appears to be one of the problems with politics and religion because politics more often than not becomes the right arm of religion to impose, control, and manipulate.

Gary Zukav: That is the history of, for example, the Catholic Church and medieval Europe. You are actually referring to a contemporary situation, but I am referring to an evolutionary shift that is changing the human species from what it has been since its origin until very recently into something that is unprecedented with a potential that did not previously exist. And that is the potential of authentic power. That is the archetype of spiritual partnership, and that is the ultimate potential of the universal human. The human that is beyond nation, beyond religion, beyond culture, beyond sex, beyond race, an individual who is a citizen of the universe first, and all else second, and whose allegiance is to life.

Rahasya: These are exciting times we find ourselves in, aren't they, Gary? I can't imagine living in any other times more exciting than living on the cusp of a collective human transformation of consciousness. This will be the ride of a lifetime for people who are detached from their fears.

Gary Zukav: Yes, yes, that's right. And this cusp is going to span several generations, and we are on the ground floor of it.

CHAPTER 8

Why We Believe What We Believe

Religion is about turning untested belief into unshakable truth through the power of institutions and the passage of time.
Gore Vidal

The Astonishing Hypothesis

Francis Krick was a physicist and he discovered the DNA molecules that transmit the genetic codes from one generation to another, a discovery that won him and James Watson the Nobel Prize. He said:

> *General ideas, especially moral ones impressed on us at an early age often become deeply embedded in our brains; it can be very difficult to change them. This may help to explain why religious beliefs persist from generation to generation, but how did such ideas originate in the first place and why do they so often turn out to be incorrect? The very nature of our brains evolved to guess at the most plausible interpretation of the limited evidence available makes it almost inevitable that without the discipline of scientific research we shall often jump to the wrong conclusions, especially about abstract matters.*

A human being is a part of the whole called by us universe, a part limited in time and space. He experiences himself, his thoughts and feeling as something separated from the rest, a kind of optical delusion of his consciousness. This delusion is a kind of prison for us, restricting us to our personal desires and to affection for a few persons nearest to us. Our task must be to free ourselves from this prison by widening our

circle of compassion to embrace all living creatures and the whole of nature in its beauty.
Albert Einstein (1879-1955)

It's a well-established fact that we are well equipped neurobiologically to fill in gaps when it comes to our senses, such as vision for instance; although we have a distinct blind spot where our optic nerve connects, our brains fill in that space without our even knowing it. This is what our brains do; it makes sense of our world based on whatever evidence it can gather and then it fills in the rest. The brain searches for patterns and does not know what to do with missing pieces, so it either fills in the gaps, or to avoid conflict, simply filters them out. In a deeper sense, this is what our brains do when we attempt to see the world and come up short of information; we fill in our blind spots with a "belief" so our world makes sense and filter out everything else as either irrelevant or false data.

We have this tendency to block out anything that contradicts our previous beliefs. Many research projects have shown that, based on belief or even gender, people neurologically block out certain concepts and even individual words. Beliefs can also cause us to hear concepts that are, in reality, not presented. We have all experienced this with friends who hear what they want to hear. What is more difficult is to see that we also do this in our personal search for truth. In fact, you may be doing it right now to some degree.

Andrew Newberg, MD, is an associate professor of radiology and psychiatry and an assistant professor of religious studies at the University of Pennsylvania. He has this to say about religious belief systems:

> *Even when their belief systems are fundamentally similar to ours, we still feel that they are significantly different. For example, Christianity, Judaism, and Islam all embrace similar notions of God, yet according to one poll nearly one-third of Americans believe that each of these religious groups worships a different deity. Even though a close investigation of the world's religions will show that the majority of human beings share similar ethical values we tend to ignore the similarities and focus on the discrepancies. Ignorance is only partly to blame. A more significant reason is that our brains are eventually prone to reject information that does not conform to our*

prior experience and knowledge. Simply put, old beliefs like habits die hard.[30]

In *Why God Won't Go Away* (2002), Dr. Newberg addresses our perception of God and other spiritual experiences by studying the brain processes that occur during meditation, prayer, and mystical experiences:

> *My research, conducted with my late colleague, Eugene d'Aquili at the University of Pennsylvania, suggests that we are naturally calibrated to have and embrace spiritual perceptions by the neurological architecture of our minds. But every individual also seems to have an abiding need to construct moral, spiritual, and scientific beliefs that explain the workings of the universe. So a belief itself is a fundamental, essential component of the human brain. As we evolved, beliefs, even superstitious ones, allowed our ancestors to make sense out of an incomprehensible, dangerous world. Their assumptions may not have been accurate, but their beliefs reduced their fears and imparted values that would facilitate group cohesiveness.*

Dr. Newberg goes on to say:

> *For those who study the nature of human consciousness, beliefs seem like a sorcerer's apprentice who is constantly playing tricks with her mind. And yet beliefs are our most important human commodity. With them we can build civilizations, make revolutions, create music and art, and determine our relationship to the cosmos. Beliefs make us fall in love, and they drive us into hate; that is why it is so crucial to understand how they work, we all have beliefs, we all need them, and they will determine humanity's fate.*
>
> *Religious and spiritual beliefs have had a particularly profound influence over human history, and yet we barely grasp how they work at a biological, behavioral, or psychological level. As a neuroscientist, I have come to realize that the study of beliefs may be the single most important quest, both scientifically and spiritually. Furthermore, I*

---

[30] From *Why We Believe What We Believe: Uncovering Our Biological Need for Meaning, Spirituality, and Truth* (2006).

think we must begin this exploration by examining the very part of us that does the believing—the human brain.

Dr. Newberg and most other neuroscientists in this line of research agree on many points and disagree on some. But the point that all of them whom I have talked with seem to agree on is that the older you get the more difficult it gets to change your beliefs, which is directly related to the way the neural pathways in our brains become static and rigid—what Dr. Rollin McCraty would call "imprints" and what Dr. Bruce Lipton might call "unconscious tapes or programs." Indeed it is a rare occasion when someone dramatically changes his or her beliefs at any age. When we are young our personalities are forming identifications and are demanding definitions, and then as we age, these neural pathways in our brains crystallize.

"The brain is a stubborn organ. Once its primary set of beliefs has been established, the brain finds it difficult to integrate opposing ideas and beliefs. This has profound consequences for individuals and society and helps to explain why some people cannot abandon destructive beliefs, be they religious, political, or psychological."

Andrew Newberg,
Born to Believe (2006)

Wherever there is a mystery in life our minds are going to go there, questioning and formulating ideas and beliefs; this is inevitable. What's important is to ask the difficult questions about our beliefs, no matter who is telling us they are true. Dialogue and questioning authority are imperative in our quest for truth. Look back at your life and you will see that we all had to let go of beliefs we had and we did so because of fresh information and maturing minds. As I mentioned before, it eventually occurs to us that Santa Claus will not fit down the chimney, and there's no way he can deliver all those packages to every child in the world in one night no matter how many elves he has, not to mention the fact that reindeer simply do not fly. But a lot of children hang on to this belief even in the face of hard facts because there's a payoff for continuing to believe this fantasy; it's called gifts in this case. And I dare say that we hold on to a lot of beliefs for this very reason, except we call them eternal life in heaven, 72 virgins, or whatever else our gift or payoff is for hanging on to outdated beliefs.

CHAPTER 9

Pascal's Wager[31]
Believing as a Form of Insurance

This particular reason for maintaining a belief with no evidence hits close to home for it is my own father's one and only reason for hanging on to his beliefs. When you first hear the argument, it sounds as if it makes sense if you refrain from giving it much thought. I recently saw a bumper sticker on a car that read, *"If you believe in Jesus and you're right, you have gained everything, if you're wrong, you have lost nothing."*

Well, I beg to differ on that last point. When we look at the price we have paid for unfounded religious beliefs in the past millennia and more, I think most rational thinking people would come to the conclusion that there is a high price we pay for believing in beliefs that are divisive, destructive, and end up being wrong. What if we used that argument for a flat world: "If you believe the world is flat and it turns out to be right, you have just

---

[31] Pascal's Wager is the application by the French philosopher, Blaise Pascal, of decision theory to the belief in God (also known as Pascal's Gambit). It was set out in the *Pensées,* a posthumous collection of notes made by Pascal toward his unfinished treatise on Christian apologetics.The wager posits that it is a better "bet" to believe that God exists than not to believe, because the expected value of believing (which Pascal assessed as infinite) is always greater than the expected value of not believing. In Pascal's assessment, it is inexcusable not to investigate this issue: *"Before entering into the proofs of the Christian religion, I find it necessary to point out the sinfulness of those men who live in indifference to the search for truth in a matter which is so important to them, and which touches them so nearly. Variations of this argument may be found in other religious philosophies, such as Islam, Hinduism, and even Buddhism."* From Wikipedia.

gained everything, but if you're wrong, you have lost nothing?" Well, true, except for the fact that most of us would have never ventured out across the oceans to discover new lands.

I think the best way to put this argument to rest would be to use the words of Sam Harris, author of *Letters to a Christian Nation* and *End of Faith* when he writes about what he calls "The Empty Wager."[32]

The Empty Wager

I have discovered that all human evil comes from this, man's being unable to sit still in a room.
Blaise Pascal (1623-1662)

Pascal suggested that religious believers are simply taking the wiser of two bets: If a believer is wrong about God, there is not much harm to him or to anyone else, and if he is right, he wins eternal happiness; if an atheist is wrong, however, he is destined for hell. Put this way, atheism seems the very picture of reckless stupidity.

But many questionable assumptions are built into this famous wager. One is the notion that people do not pay a terrible price for religious faith. It seems worth remembering in this context just what sort of costs, great and small, we are incurring on account of religion. With destructive technology now spreading throughout the world with 21st-century efficiency, what is the social cost of millions of Muslims believing in the metaphysics of martyrdom? Who would like to put a price on the heartfelt religious differences that the Sunni and the Shia are now expressing in Iraq (with car bombs and power tools)? What is the net effect of so many Jewish settlers believing that the creator of the universe promised them a patch of desert on the Mediterranean? What have been the psychological costs imposed by Christianity's anxiety about sex these last 70 generations? The current costs of religion are incalculable. And they are excruciating.

While Pascal deserves his reputation as a brilliant mathematician, his wager was never more than a cute (and false) analogy. Like many cute ideas in philosophy, it is easily remembered and often repeated, and this has lent it an undeserved air of profundity. If the wager were valid, it

32 Reprinted with permission from the Reason Project.

could be used to justify any belief system (no matter how ludicrous) as a "good bet." Muslims could use it to support the claim that Jesus was not divine (the Quran states that anyone who believes in the divinity of Jesus will wind up in hell); Buddhists could use it to support the doctrine of karma and rebirth; and the editors of *Time* could use it to persuade the world that anyone who reads *Newsweek* is destined for a fiery damnation.

But the greatest problem with the wager—and it is a problem that infects religious thinking generally—is its suggestion that a rational person can knowingly will himself to believe a proposition for which he has no evidence. A person can profess any creed he likes, of course, but to really believe something, he must also believe that the belief under consideration is true. To believe that there is a God, for instance, is to believe that you are not just fooling yourself; it is to believe that you stand in some relation to God's existence such that, if he didn't exist, you would not believe in him. How does Pascal's wager fit into this scheme? It does not.

Beliefs are not like clothing: Comfort, utility, and attractiveness cannot be one's conscious criteria for acquiring them. It is true that people often believe things for bad reasons—self-deception, wishful thinking, and a wide variety of other cognitive biases really do cloud our thinking—but bad reasons tend to work only when they are unrecognized. Pascal's wager suggests that a rational person can knowingly believe a proposition purely out of concern for his future gratification. I suspect no one ever acquires his religious beliefs in this way (Pascal certainly didn't). But even if some people do, who could be so foolish as to think that such beliefs are likely to be true?

CHAPTER 10

Spirituality vs. Religion

Some of you may believe that there is very little conflict between religion and spirituality so we should clear that up before we move on. You will often hear someone say, "I consider myself a spiritual person, not a religious person." What follows are some of the differences between spirituality and religion.

- Spirituality says that God is within us and that we don't need anyone else, such as rabbis and priests, to make that divine connection for us.
- Religion says that we are separate from a God who is outside of us and that we need someone to intervene on our behalf to make the divine connection.

- Spirituality says that we are free to make choices on our own and that we must take personal responsibility for our actions.
- Religion says that we have no choice but to obey and follow blindly with faith.

- The whole thrust of spirituality is based on love and oneness.
- Religion is riddled with the idea of shame, guilt, fear, and suffering and unworthiness.

- Spirituality does not require us to tithe or make donations.
- Religion requires donations and most say your reward will be waiting in heaven when you die.

- Spirituality says there is no hell or judgment awaiting us from a vengeful jealous God and that we are all loved unconditionally.

- Religion says there is a hell and we will be judged by a very capricious God based on how much we pleased him.

- Spirituality says that we are free to choose one of many paths to God.
- Most religions say they are the only way and all others will lead you to eternal suffering in hell.

- Spirituality says that we should not be ashamed of our sexuality and in fact it is a sacred act to be celebrated.
- Religion has taught us to be ashamed of our sexuality for the past 70 generations or more.

- Spirituality teaches us that we are caretakers of the earth that we live on.
- Religion tells us to subdue the earth and have dominion over all living things.

- Spirituality teaches that God is within every one of us.
- Religion puts God in a heavenly abode somewhere out there.

- Spirituality says we are born in innocence.
- Religion says we are born in sin, and we remain that way.

- Spirituality teaches unconditional love.
- Religion teaches that even God has conditions for you to be loved.

- Spirituality teaches us that we should love and respect each other unconditionally.
- Religion has given us the foundation for wiping out or attempting to enslave whole cultures of people.

- Spirituality teaches us that we have the answers if we only look.
- Religion teaches us that we are lost and only it has the answers.

So the question is: *"Which one feels better to you and makes more sense in creating a sustainable world?"*

What is the role of the established religions in the arising of the new consciousness? Many people are already aware of the difference between spirituality and religion. They realize that having a belief system, a set of thoughts that are regarded as an absolute truth, does not make you spiritual no matter what the nature of those beliefs are. In fact, the more you make your thoughts and your beliefs into your identity the more cut off you are from the spiritual dimension within yourself. Many religious people are stuck at that level. They equate truth with thought, and as they are completely identified with thought and their minds, they claim to be in sole possession of the truth in an unconscious attempt to protect their identities. They don't realize the limitations of thought and unless you believe and think exactly as they do you are wrong in their eyes. And in the not-too-distant past they would have felt justified in killing you for not sharing their beliefs, and many still do, even today.

The new spirituality, the transformation of consciousness, is arising to a large extent outside the structure of the existing institutionalized religions. There were always pockets of spirituality, even in mind-dominated religions, although the institutionalized hierarchies felt threatened by them and often tried to suppress them. A large-scale opening of spirituality outside the religious structures is an entirely new development. In the past this would have been inconceivable, especially in the West, the most mind dominated of all cultures and where the Christian Church had a virtual franchise on spirituality. You could not just stand up and give a spiritual talk or publish a spiritual book unless you were sanctioned by the Church; and, if you were not, it would quickly silence you. But now even within certain charters and religions, there are signs of change. It is heartwarming and one is grateful for even the slightest signs of openness, such as Pope John Paul II's visiting a mosque and a synagogue. Partly as a result of the spiritual teachings that have arisen outside the established religions, but also because of an influx of the ancient Eastern wisdom teachings, a growing number of followers of traditional religions are able to let go of identification with form, dogma, and rigid belief systems and to discover the original depths that are hidden within their own spiritual traditions at the same time they discover the depths within themselves. They realize that how spiritual one is has nothing to do with belief and everything to do with our state of consciousness. This in turn determines how we act in the world and interact with others.

Those unable to look beyond form become even more deeply entrenched in their beliefs, that is to say, in their minds. We're witnessing not only an unprecedented influx of consciousness now, but also an entrenchment and intensification of the ego. Some religious institutions will be open to the new consciousness, while others will harden their doctrinal positions and become part of all the other man-made structures through which the collective ego will defend itself and fight back. Some churches, cults, and sects or religious movements are basically collective egoic entities. Those entities are as rigidly identified with their mental positions as the followers of any political ideology that is closed to any alternative interpretation of reality.

But the ego is destined to dissolve and all its created structures, whether they be religious or other institutions, such as corporations or governments, will disintegrate from within no matter how deeply entrenched they appear to be. The most rigid structures and those most impervious to change will collapse first. This has already happened in the case of Soviet communism. How deeply entrenched, how solid and monolithic it appeared, and yet within a few years it disintegrated from within; no one foresaw this. All were taken by surprise. Many more such surprises are in store for us, one of which may be what seems to be happening in the United States. We show all the signs of the rise and fall of a civilization. Historically, one of the signs of a decline is the belief that we can expand and dominate a large part of the world with our culture and beliefs. This is impossible, yet we show every sign of attempting to bring it about.

The other sign of the decline of a civilization is national debt, and the United States has more debt than it can ever repay. The truth be known, it isn't expected to repay it; it's only meant to keep us in poverty as financial slaves. Most of our debt is interest on preexisting debt. The only way this can go on is if the citizens continue to believe what is told to them. My recommendation is to not believe anything the government tells you; look into what's been going on, think for yourself, and take action with any number of groups that are forming to bring light to this and many other situations. The government knows that it's a simple step to take a person who subscribes to Iron Age beliefs and turn him into a patriot who will kill and die for his country, all the while believing that the Federal Reserve is federal and that there are reserves, neither of which is true. I have not met one "spiritual" person who believes what the government or religious organizations say without checking it out, while on the other hand, most religious people openly and courageously believe every word without question. If someone does question these things he or she is viewed as unpatriotic and losing faith in God, and in the United States these groups are linked.

CHAPTER 11

The Power of Beliefs

For those who may be thinking, "What's the big deal with believing something?"[33] Here's the big deal; beliefs are powerful and run deep into our psyches. To illustrate the power of a belief, imagine for a moment that someone came to your home and said, "I just saw your child get run over and killed by a drunken driver." Be assured that if you believe that to be true, it will affect your actions, your values, your thoughts, your judgments, and what you will be doing in the next moments of your life. Now imagine the power a religious belief has over you when you take it to the level of eternal punishment or pleasure for not only your actions, but also for your thoughts and emotions.

I could write another book on this but showing the number of insane and absurd things we believe is not my purpose; showing the *neurological* and *social* consequences is my purpose, but it's good to see a few beliefs that are held in the collective consciousness of billions of people on this planet. To put this in even better perspective, imagine that only one person, not billions, would have and practice a major religious belief. Then you can readily see the absurdity and dangerousness of certain religious and nationalistic beliefs, not to mention the fact that the person would be immediately institutionalized. On our planet, if you have an absurd and even dangerous belief, you might face the prospect of being shunned or alienated from society. But if you can get even a few hundred thousand people to believe

[33] Don't get this type of believing confused with something that would be better to call "focused attention or intention," in which we are using the power of thought, but even here it must be understood that even focused attention or intention needs to be validated along the way by rigid scientific research by unbiased researchers using control groups.

you, you become a religious leader or a political revolutionary. Along this same line of reasoning, if you kill one person, you are a murderer, but if you manage to kill a million people, you are a conqueror and will be honored as a hero of the people you killed for.

Here are just a few beliefs that people believe and have believed on this planet of ours:

- If you say the right words from a language that's all but dead, over a small cracker, you will miraculously change it into the flesh of a man who was born from a virgin 2,000 years ago, who was resurrected and brought back to life three days later, who left this planet into the clouds, and who will come back some day and punish all those who don't believe this for eternity by banishment to hell.
- Hindus disagree over whether a Shiva or Vishnu is a higher lord, and many have been killed for their belief in this matter.
- If you die as a martyr while killing infidels (people who do not believe as you do), you will spend eternity in heaven with 72 virgins by your bedside to have an eternity of orgasms.
- The Mormons say that the Garden of Eden is in Missouri.
- In America, 24 percent of the people believe Jesus will come back and save them, and only them, within the next 50 years, and this will happen as soon as we can fulfill the prophecies of the Bible, which calls for Armageddon, our waters to be undrinkable, our oceans dead, and our forests knocked down.[34] [35] These people vote and some of them hold office and make important decisions on our behalf that have a direct effect on our lives.

---

[34] Armageddon is the final climactic battle between God and Satan the devil, written in the Book of Revelation in the Christian New Testament, or more generally, an apocalyptic catastrophe.

[35] 2 Thessalonians 1:6-9: God deems it just to repay with affliction those who afflict you . . . when the Lord Jesus is revealed from Heaven with his mighty angels in flaming fire, inflicting vengeance upon those who do not know God and upon those who do not obey the gospel of our Lord Jesus. They shall suffer the punishment of eternal destruction and exclusion from the presence of the Lord and from the glory of his might.

- In America 62 percent want creationism taught in schools, and 44 percent want it exclusively taught, even though there is absolutely no supportive scientific evidence for creationism and almost daily we accumulate more scientific evidence in support of evolution.
- In America 44 percent literally believe the God of Moses, in some sort of cosmic real estate deal, gave a piece of land to the children of Abraham because they were the chosen people and favored by God-well, not all the children of Abraham, just the ones by Abraham's wife Sarah. The other children, out of wedlock, were the Arabs and have nothing coming, even though recent genetic testing shows there is absolutely no difference in the two races.
- Women are hardly more than an afterthought by God and were created to keep men comfortable and be at their service; this notion is spelled out in great detail in the Bible and the Quran.
- Honor killings are still practiced in many Arab nations. An honor killing could result from your daughter's coming home and informing you she was raped and you, as a Muslim, would be honor bound to kill her. This is a good example of how social evolution has made it all but impossible for most world religions to obey their Holy Books to the letter of the law. And if you're a Christian and your wife is raped and she didn't scream loudly enough for someone to hear, according to the Bible you can beat her or stone her to death.
- It was not uncommon in many ancient cultures for a child to be born for the explicit purpose of raising it to be butchered or burned alive as a sacrifice to the God of the day to keep the sun in place.
- In ancient Rome, sometimes children were slaughtered so the future could be read in their intestines.
- If you wanted a date with a Dyak woman living in Borneo, you would need to have a net full of human heads as a "love" offering.
- On a Fijian island there was a sacrament called Vakatoga, in which the victim's limbs were cut off and eaten in front of him.
- The Iroquois would take captives from other tribes and hold them for years and they would often marry and have children, but their fate was sealed because at some point the captives and their children would be flayed as an offering to the God of war.
- There are a variety of beliefs in the Bible that would no longer be tolerated in today's world, yet millions of people hold the Bible

and say proudly that it is the immutable word of the creator of the universe and is to be taken literally.[36]

Keep in mind that all these beliefs are and have been religious beliefs and center around what different groups of people thought they knew about an invisible God. It might be worthwhile to mention that most of these groups knew nothing or very little about any of the sciences we have today.

It was in ancient history that today's world religions have their roots and it was in a climate of superstition, fear, and hardly any physical science. It's only in this type of atmosphere that a society could possibly come up with an idea that the "loving" God and creator of this universe waited until 2,000 years ago to send his only "beloved" son to be sacrificed for all the sins and misbehavior of others. This is such a blatant and obvious carryover of the scapegoating ritual that was going on at the time that I hardly know where to start or stop when writing of the absurdity of billions of people still believing this idea in today's world.

As I mentioned before, I could write another complete book on the crazy things we believe but I'm sure you get the point. And the point isn't if it's true or not true; the point is that there's absolutely no possible way to know

---

[36] Leviticus 25:44-46: As for your male and female slaves whom you may have: you may buy male and female slaves from among the nations that are around about you. You may also buy from among the strangers who sojourn with you and their families that are with you, who have been born in your land; and they may be your property. You may bequeath them to your sons to inherit as a possession forever; you may make slaves of them, but over your brother in the people of Israel you shall not rule, one over another, with harshness.

Exodus 21:7-11: When a man sells his daughter as a slave she shall not go out as the male slaves do. If she does not please her master, who has designated her for himself, then he shall let her be redeemed; he shall have no right to sell her to a foreign people, since he has dealt baselessly with her. If he designates her for his son, he shall deal with terror as with a daughter. If he takes another wife to himself, he shall not diminish her food, her clothing, or her marital rights. And if he does not do these three things for her, she shall go out for nothing, without payment of money.

I have no desire to get into a long dissertation about all the absurdities to be found in the Bible, but I believe it's wise to keep in mind the many passages that have led people and whole cultures to commit atrocities throughout history.

if Mary, who lived 2,000 years ago, was a virgin or not; I doubt I would have been able to know for sure even if I had been there. As Sam Harris once said, *"If someone says they know that Jesus was born of a virgin, they are lying."* Indeed, it is impossible to know if any of the former statements are true, yet they are believed with such fervor that to suggest otherwise would be to put yourself in harm's way with most believers. Now remember when we thought about how we would feel and think if someone told us that our child was run over by a drunken driver? How do you think these types of beliefs affect our actions and thoughts? The whole world is being affected and infected by these beliefs right now. Turn on the news and see for yourself! Billions of people worldwide maintain and organize their entire lives, and armies, around absurd and dangerous beliefs such as these.

Our irrational beliefs are causing untold suffering in the world in thousands of ways. Think of the millions who have died and will continue to die in sub-Saharan Africa of AIDS because religious organizations believe it's wrong to use condoms, even in marriage. This is going on while billions of people in the world are fully aware that this rule was originally meant to encourage the reproductive advantages of maintaining a high birth rate, thereby ensuring that the belief will continue. Of course the goal isn't the spread of AIDS but believing blindly doesn't take that into account. This is the logical consequence of a belief that is irrational and lacks any empirical evidence for its justification.[37] Of course to consider the implication that genocide is behind this decision is even darker.

This excerpt is from the *LA Weekly,* June 24, 2004.

> Lethal new regulations from President Bush's Centers for Disease Control and Prevention (CDC) in Atlanta, quietly issued with no

[37] In this day and age it's difficult to believe that the Vatican still opposes condom use, even to prevent the spread of HIV, although the pope is rumored to be reconsidering his policy on this matter. This is a good example of something once held by faith and belief that will someday be changed by rational investigation. This has happened time and again throughout history to dogmas that come into direct conflict with rational and empirical evidence. There's only one other explanation for the pope to change his beliefs, by the way, and that is that he has a direct phone line to God and only he can talk to him or her. I think most of us no longer accept this.

fanfare last week, complete the right-wing Republicans' goal of gutting HIV-prevention education in the United States. In place of effective, disease-preventing safe-sex education, little will soon remain except failed programs that denounce condom use, while teaching abstinence as the only way to prevent the spread of AIDS. And those abstinence-only programs, researchers say, actually increase the risk of contracting AIDS and other sexually transmitted diseases (STDs).

Published on June 16 in the Federal Register, the censorious new CDC guidelines will be mandatory for any organization that does HIV-prevention work and also receives federal funds—whether or not any federal money is directly spent on their programs designed to fight the spread of the epidemic. (The CDC is the principal federal funder of prevention education about HIV and AIDS, and its head a Bush appointee). It's all couched in arcane bureaucratese, but this is the Bush administration's Big Stick—do exactly as we say, or lose your federal funding. And nearly all of the some 3,800 AIDS service organizations (ASOs) that do the bulk of HIV-prevention education receive at least part of their budget from federal dollars. Without that money, they would have to slash programs or even close their doors.

These new regulations require the censoring of any "content"—including "pamphlets, brochures, fliers, curricula," "audiovisual materials" and "pictorials (for example, posters and similar educational materials using photographs, slides, drawings or paintings)," as well as "advertising" and Web-based info. They require all such "content" to eliminate anything even vaguely "sexually suggestive" or "obscene"—like teaching how to use a condom correctly by putting it on a dildo, or even a cucumber. And they demand that all such materials include information on the "lack of effectiveness of condom use" in preventing the spread of HIV and other STDs—in other words, the Bush administration wants AIDS fighters to tell people: Condoms don't work. This demented exigency flies in the face of every competent medical body's judgment that, in the absence of an HIV-preventing vaccine, the condom is the single most effective tool available to protect someone from getting or spreading the AIDS virus.

This type of political nonsense would not be tolerated or even possible if it weren't for deeply held religious beliefs about sex being dirty; remember, it was the original "sin." The head of the CDC even went on to say that if a cure for AIDS came out he would be against it because it would encourage promiscuous sex. In March 2009 the pope announced that the use of condoms would actually make the AIDS epidemic even worse although because of public opinion he is reconsidering this view.

We could also talk about stem cell research and I don't know anyone who has put it more plainly than Sam Harris.[38] The point he brings up is that the reasoning behind stopping stem cell research is to alleviate any possibility of suffering that the "potential" human being may go through, plus the fact that it is a potential human being with a soul. First of all, anyone who puts the interests of a blastocyst before the suffering of a child with a spinal injury needs to do some deep reflection on his or her metaphysical reasoning for this. And if it's the argument that it's a potential human being, then we need to consider what we have learned from our genetic research, which tells us that every cell in our bodies is a potential human being and none of us thinks twice about the millions of cells we wash off every day down the drain. And if it's the idea of a soul, then what do we do with the cell that splits and becomes two zygotes—does it now have one-half of a soul? Anyway you look at it or calculate it, it becomes absurd at some point.

If that's not enough, we merely need to think about the small child with a crippling disease that keeps him or her from walking or our elderly father or mother with Alzheimer's or Parkinson's disease. Stem cell research is the most promising medical research ever and it's being blocked in Congress because of people's religious beliefs that have absolutely no substance, reason, or logic. These are the types of social consequences that are extremely troubling for some people in our society.

None of this is to say that we should give free reign to scientists; I can see right away that we need oversight committees to safeguard our research efforts. And this brings up another problem in this respect; by making things such as birth control and stem cell research against the law do you think for a moment that they will not happen and aren't happening at this very moment? We drive researchers underground, where we can no longer have access to what they are doing and that is also dangerous. We force young pregnant girls down dark alleys to have abortions. We've tried in the past to make laws against things such as alcohol and created the world's most notorious gangsters ever and made it possible for them to earn enormous amounts of money and enter into high government positions. Why? Religious values

---

38 *Letters to a Christian Nation*, 2006.

were imposed on society by government agencies for purely religious reasons for which we all pay a price now and for many generations to come. Here is an example of the narrow-mindedness in the United States: The Rev. Jay Scott Newman, a Catholic priest, wrote this in a letter to parishioners at St. Mary's Catholic Church in Greenville, South Carolina, condemning whomever voted for Barack Obama. "*Voting for a pro-abortion politician when a plausible pro-life alternative exists constitutes material cooperation with intrinsic evil, and those Catholics who do so place themselves outside of the full communion of Christ's Church and under the judgment of divine law.*"

As long as we are covering some of our beliefs, here are some statistics taken in the United States on which our social beliefs have a direct effect:

In a University of Minnesota study, more than 2,000 people were asked which of their fellow citizens lacked the proper "vision of American Society." More than blacks, gays, immigrants, lesbians, or Muslims, atheists are viewed as least American, according to the survey. More than 36 million atheists in the United States do not believe in God. The First Commandment is: *Thou shalt have no other gods before me.* Deuteronomy 17: 2-7 tells us that we should take anyone who breaks this commandment and stone him or her to death. Deuteronomy 13:13-19 tells us that if we find a town that does not believe in God we are to burn it to the ground and kill all the livestock and inhabitants, although some special concessions are made for young virgin girls.

Christians make up 75 percent of the population and also 75 percent of the prison population, according to the Federal Bureau of Prisons. Atheists make up 10 percent of the population and make up only 0.2 percent of the prison population.

We need to keep in mind that in most cases atheists are simply stating the obvious fact that they don't know. If you run into an atheist who says he or she knows there is no God, you're simply dealing with the other side of the belief of those saying that they know God does exist.

So Who Is the Fool?

Here's a statement by George Bush Sr.: "*No, I don't know that atheists should be considered as citizens, nor should they be considered as patriots. This is one nation under God.*" This type of belief is backed up in the Bible in verses such as Psalms 14:1, where it says, "*The fool has said in his heart, 'There is no God.' They are corrupt, their deeds are vile; there is none who does good.*"

So if, according to scripture, a person who questions whether or not there is a God is a fool, let's take a look at some of the fools in our culture and you can decide for yourself.

- Isaac Asimov, a fool and atheist, was a brilliant biochemist and prolific and almost prophetic science fiction writer. He also did much to popularize science for the public mind.
- Noam Chomsky, another fool who is a professor of linguistics at MIT, formulated the theory of generative grammar and was considered the leading American public intellectual of 2005. Between the years of 1980 and 1982 he was cited as a source more often than any other living scholar.
- Francis Crick, but another fool and atheist, codiscovered the structure of DNA and won the 1962 Nobel Peace Prize for physiology or medicine.
- Marie Curie was a fool who won the 1903 Nobel Peace Prize for physics and the 1911 Nobel Peace Prize for chemistry.
- Richard Dawkins, the fool and atheist extraordinaire, is an Oxford professor of biology and zoology.
- Daniel Dennett, another fool, is a leading philosopher of the mind, cognitive scientist, and evolutionary biologist.
- Thomas Edison, yes, another atheist and fool, invented the incandescent light bulb and was instrumental in a number of other inventions that changed the world.
- Stephen Gould, a professor at Harvard University, was a leading paleontologist and evolutionary biologist.
- Massimo Pigliucci is the fool who has PhDs in botany, genetics, and philosophy.
- Steven Pinker is the fool who is a professor of psychology at Harvard, the director of the Center for Cognitive Neuroscience at MIT, and one of *Time* magazine's top 100 most influential people.
- Karl Popper is yet another fool who is considered the most influential philosopher of science in the 20th century.
- Carl Sagan—I'm sorry, I just can't bring myself to call him a fool even jokingly. He was a world-famous astronomer, astrobiologist, and he did much to popularize science for old and young minds. He was voted as one of the 100 greatest Americans by the Discovery Channel. He was also a 1978 Pulitzer Prize winner, and he was the

creator of the still-popular *Cosmos* series that has opened the minds of millions of people around the world.

- Michael Shermer is the fool who is an experimental psychologist and historian of science and written numerous books. He founded *Skeptic* magazine, writes for *Scientific American,* and is considered America's leading skeptic.
- James Watson is the fool who codiscovered DNA and won the 1962 Nobel Peace Prize for physiology or medicine.
- E. O. Wilson, the fool who was a professor of entomology and evolutionary biology at Harvard, was named one of America's top 25 most influential people and also won two Pulitzer Prizes along with the Crafoord Prize.
- Warren Buffett is a fool who donated 37 billion dollars to charity.
- Bill Gates foolishly donated another 30 billion dollars.
- Ernest Hemingway is the fool who was a Nobel Peace Prize winner and Pulitzer Prize winner.
- Mark Twain—of course we all know this fool.

It's obvious that I could go on and on listing the people who would be classified as fools because they don't hold to the traditional religious beliefs of the day. But without these "fools," where would we be? It has been the great minds throughout history who have held the hope of mankind; many were persecuted in their own time only to be acknowledged after their death or persecution.

Those who wish to appear wise among fools,
among the wise seem foolish.
Quintilian (c. 35-c. 100)
De Institutione Oratoria

So what can we do? We can wake up and learn from history and stop killing and persecuting people who are speaking out in our time with a clear voice about the insanity of believing this type of nonsense. If we don't learn from history, what do you think will happen? Yes, we will repeat it and our future generations, if we don't make the world uninhabitable, will look back the way we look back and wonder how we could have been so blind, ignorant, and barbaric to continue the perpetuation of belief systems that almost destroyed the world in the face of hard evidence that they were based on a faulty premise from the beginning.

CHAPTER 12

Dhara Lemos, Neale Donald Walsch, and Rahasya

Time for Action
A Talk with Neale Donald Walsch[39]

When my wife and I visited Neale Donald Walsch at his home in Ashland, Oregon, we sensed an urgency in his tone regarding

[39] Neale Donald Walsch is a modern-day spiritual messenger whose words continue to touch the world in profound ways. With an early interest in religion and a deeply felt connection to spirituality, Neale spent the majority of his life thriving professionally, yet searching for spiritual meaning, before beginning his now-famous conversation with God. His *With God* series of books has been translated into 34 languages, touching millions of lives and inspiring important changes in their day-to-day lives. For more information visit www.nealedonaldwalsch.com.

how we have held onto outdated religious beliefs. Neale wrote *Conversations with God* a few years back (1996) and has touched the hearts of millions of readers. I think what I admired most about him was his freedom and courage to speak his mind like a man who had nothing to lose. It takes a lifetime of asking tough questions that no one wants to hear to get to where he is at this point in his life. It also takes real-life experiences to see what is important in life and Neale has had many such experiences. I urge you to see his movie of the same title that documents his life as a homeless person who finds what he is looking for by looking inside and listening.

Of course we again find ourselves in a delicate area when we start listening to voices in our heads. This is where we need to be open to possibilities and at the same time use a little common sense. I think it's a good rule of thumb to look at the quality of the information coming to us and use caution when defining its source. That being said, I must admit that Neale presents a strong case and his definition of God is far beyond what most traditional religious organizations could or would embrace.

Rahasya: If you had to define God, what would your definition be?

Neale Donald Walsch: God is everything. Everything seen and unseen. Physical and nonphysical. And the sum total of all that is reducible to a single word: *energy*. What we see in life in its many forms are manifestations of that energy or vibrations. The basic essence of the universe vibrating at different frequencies to produce the physical and nonphysical, depending on the rate of speed. So we're talking about the primal energy, a primal source of "what is." In the East, it is sometimes called "chi." And that energy experiences itself. We are that energy, and we can experience the energy. Even more astonishingly, we can manipulate that energy. And depending on the level of mastery to which we have obtained, we can control it to an even greater degree. True masters have been able to control it utterly.

Rahasya: Do you think our destiny is to play more of a role in being conscious cocreators?

NDW: We are cocreators. It is our destiny to accept and ultimately to realize that. Masters, far more articulate than I, have called that self-realization. And so the question is, "What will cause it to occur and when?"

Rahasya: I heard Carl Sagan say once that the reason there might not be a lot of planets out there communicating with us is because most planets don't make it through their technological adolescence. And I think that's where we're at right now; we're going through a technological adolescence. A little like a young child with dangerous weapons.

NDW: Exactly. We're going through a technological adolescence and we're going through a theological infancy. The difficulty in today's world is our technology and science has outrun our theological advances. The reason for that is in technology and science, we have had the courage to ask the single question that theology has been afraid to ask, "Is it possible that there's something I don't know about this, the knowing of which would change everything?"

Rahasya: And what's amazing is that history has shown us that truth evolves and changes over time, and when truth doesn't evolve history has also shown us the atrocities that are created by the suppression of truth.

NDW: Of course, because life itself is evolving. Even when you think you have the final answer, life will fool you because life will say, "Well, you think you've caught up with me, do you?" And so when we have what appears by all measure and means to be the final answer of something, the universe plays a trick on us. Scientists thought they understood the physical universe and then they discovered quantum physics. The genius of Einstein causes us to wonder what he saw that we didn't see until years later. So it's the same thing in theology, and the visionaries of our time are people who are the Einsteins of theology. That is, they're seeing things that it'll take the rest of the world 50 years, or 100 years, to see. But what must occur for humanity to evolve, to get past the infancy in this theological experience, is for the human species to ask the unaskable question, and to question its prior assumption. We have been prohibited by the religions of the world. If scientists, medical researchers, or technology students were prohibited from questioning prior assumptions, we would be nowhere. All advances have been made through that magic of having the courage to question the prior assumptions we have made. We have not had that courage in theology. What if everything we thought about God up to this point was not necessarily true? Tomorrow's God is what emerges from our willingness to question our prior assumption. What if some of it were true, but not all of it? For that matter, if you just changed one aspect of the equation, all of the rest of the numbers on the blackboard are changed, as Einstein found out.

Rahasya: I believe you just hit on the reason they don't want to question their original assumptions about God. There are a lot of people on this planet who have a vested interest in maintaining the prior assumption and again, history has shown what happened to the people who questioned those assumptions. To tell you the truth, when I was eight years old I questioned those assumptions and never quit.

NDW: I need to be honest with you—most people have. I did along with a lot of others. But our society beats that questioning out of us: "Don't you dare question it." And so all but the most courageous among us have let go of their childhood awareness that we don't have all the answers. And the reason that children know that and adults don't is that children have just come from the other realm, the realm of spirituality.

Rahasya: So what I'm wondering is what do you think is behind some of the films being released, such as* What the Bleep, Indigo, Indigo Evolution, The Da Vinci Code, The Peaceful Warrior, *and now your movie coming out. Why do you think people are opening up to this?

NDW: I think that humanity is losing patience with itself. And out of that loss of patience we are seeking new ways to reveal ourselves to ourselves. And academia and education can only go so far. And the belief systems of humanity, with regard to God, and life, and each other, are so deeply ingrained in us that it's very difficult to teach us, through standard educational pedagogy, a new idea. So society collectively has to do what mechanics would call a workaround. And the workaround is entertainment, the great spiritual workaround, and entertainment provides us an arena. People are beginning to collectively gather around an organizing principle that reminds them of what they've always known. It's a death to who we "think" we are but that death is a birth into the ultimate experience of who we really are.

Rahasya: I really think that if somehow I could get inside your consciousness, and experience life through your eyes and ears, and be in your situation, I would have a different set of memories because I have a different set of experiences. But when I say, "I am," I really feel it's the same "I am" as when I say it from this point of view.

NDW: Of course you would have the identical awareness, different memories leading to an identical awareness. We are all one.

Rahasya: As a final thought, what would you want to say to anyone reading this?

NDW: Well, I think the time has come to stop playing in our minds with these ideas and begin to start doing something in the physical world. The message is we create our own reality but to do that we need to go out into the world and "do" something . . . anything. If we don't do that in these next few decades, I think that life as we know it will no longer exist on this planet. I frankly am not frightened by that prospect or concerned about it for myself because I'm an old guy now and I'm very clear about the fact that I won't even be here during those tumultuous times should we not do what I am suggesting. I do feel saddened and disappointed on a human level for my children and my children's children. I consider myself the last generation to experience life as we now know it, unless we make these choices and undertake these actions of which I speak. Now what I think we have to do in these years ahead is to take action on the ground, in physicality, not just talk about it; that would allow humanity to alter its cultural story. And that could create the sense of possibility for a new spirituality to emerge upon the earth. To actually undertake what I call in my poetic moments a civil rights movement for the soul. The last great civil rights movement of humanity. Freeing human beings at last from the oppression of their beliefs in a violent, angry, and vindictive God. And to undertake this civil rights movement for the soul, to create the space of possibility for that new spirituality to emerge, we need to organize and gather ourselves together in large numbers that we might exponentially increase the power of individual attempt, to magnify it, to globally impact in proportion.

CHAPTER 13

Why Our Founding Fathers Separated Church and State

Since we are getting into some delicate and highly charged areas of thought I suppose I should say a few more words about myself. I should clarify the fact that I am not a Christian, Muslim, Jew, Republican, Democrat, nor a member of any other religion or political movement or ideology. I do see myself as a very spiritual and moral person, although I dislike using the word "moral" because of the misused social definitions that have been associated with it. I do have some beliefs about our existence and the world we live in, but I am very cautious not to mix what I "believe may be true" with what I "know to be true" or simply wish were true because it makes me feel good to believe it. For the record, I do believe there is some kind of organizing force that one could refer to as God, although I don't believe "it" is a personal God who listens to prayers and sometimes decides to help and other times not. I simply cannot envision a God who is self-aware and watches from afar with unsympathetic detachment. However, I am totally open to further enquiry and experiences that may change my thoughts on this.

Shake off all the fears of servile prejudices, under which weak minds are servilely crouched. Fix reason firmly in her seat, and call on her tribunal for every fact, every opinion. Question with boldness even the existence of a God; because, if there be one, he must more approve of the homage of reason than that of blindfolded fear.
Thomas Jefferson (1743-1826)

I'm not affiliated with any belief system so I have a freedom that a lot of people may not have and this turns out to be very important. Einstein once said that it would be highly improbable to come up with a solution to a problem from the same level of consciousness that created the problem in the first place, and yes, I do see beliefs, and religious beliefs in particular, as being problematic in today's world, especially when they enter into the realm of politics like organized religions do, which is what we're seeing today in the United States and Middle East. Then add to that high technology, and I say we have a problem in the making. What was once only probable is quickly becoming inevitable if we continue on our present course. I would like to add that I think it's very important to respect other people's beliefs, but like the respect we give any other aspect of society, it is *earned* through open dialogue and intellectual honesty and the ability to be questioned as to its authority and validity. This is an important point because we never fail to question scientific discoveries and research, but for some reason we fail to question a person's belief about God and his or her religious doctrines. Of course the reason for this is that religious beliefs are based on faith and you can't rationally and intellectually debate a faith-based belief—you either believe it or you don't.

They were so strong in their beliefs that there came a time when it hardly mattered what exactly those beliefs were; they all fused into a single stubbornness.
Louise Erdrich

Imagine for a moment the reception a biologist would get if he or she were to come out with a new theory of life that had absolutely no empirical or historical evidence to support it, no verifiable research, and no personal experience of its truth; and when he or she was asked why he or she believed this theory to be true all he or she had to say was, "Well, because it gives my life meaning and makes me feel good by believing it's true" or "I heard an inner voice telling me it was true." Obviously he or she would be ignored by his or her peers, not because the theory was absurd, but because the basis for thinking it was true was absurd. Religion is the only branch of knowledge that has been granted special dispensation from the need to validate its beliefs, which, it claims, are established truths through faith. This needs to change and change it will.

As we observe what's going on, and what's been going on for thousands of years in our world, it's easy to come to the conclusion that we simply do not have a solution that will work for everyone on the planet as long as there are vast groups of people who still believe they, above all others, have the truth and represent the one true God. This is egocentric thinking and narcissism at its best, or I should say, worst. Every solution we have comes from the same level of awareness that created the problem in the first place, and indeed, most of our solutions end up exacerbating the very problem we're trying to solve. We live in a self-imposed prison created by the illusion that we are separate beings living on a planet, and that we have the right to have dominion over it because we are superior and chosen by a God to do so. This divisive belief alone is responsible for so many of our environmental problems today because we see ourselves as separate from the very world in which we live, breathe, and have our being. Imagine for a moment the effect on our bodies if our liver cells started thinking they were somehow elite among the other cells and stopped carrying out their function at the expense of the rest of the body; we call this cancer, by the way.

When I do good, I feel good; when I do bad,
I feel bad, and that is my religion.
Abraham Lincoln (1809-1865)

Bill Moyers has been a longtime hero of mine because of his dedication to truth, no matter where it leads.[40] What follows is from an acceptance speech he gave at Harvard's Global Environmental Citizen awards ceremony

---

[40] Just so there's no mistake over the integrity involved in what Bill Moyers might have to say, he is the recipient of the 2006 Lifetime Emmy. "Bill Moyers has devoted his lifetime to the exploration of the major issues and ideas of our time and our country, giving television viewers an informed perspective on political and societal concerns," according to the official announcement, which also noted, "The scope of and quality of his broadcasts have been honored time and again. It is fitting that the National Television Academy honored him with their highest honor—the "Lifetime Achievement Award." He has received well over 30 Emmys and virtually every other major television journalism prize, including a gold baton from the Dupont Journalism awards, a lifetime Peabody award, and a George Polk Career Award (his third George Polk Award) for contributions to journalistic integrity and investigative reporting.

on December 1, 2004. I include his description here because it shows the serious implications of having religious beliefs so deeply entrenched in our government and the *social consequences* of such a situation.[41]

> The most unmanageable [situation] of all . . . could be the accelerating deterioration of the environment, creating perils with huge momentum like the greenhouse effect that is causing the melt of the Arctic to release so much freshwater into the North Atlantic that even the Pentagon is growing alarmed that a weakening Gulf Stream could yield abrupt and overwhelming changes, the kind of changes that could radically alter civilizations.
>
> That's one challenge we journalists face—how to tell such a story without coming across as Cassandras, without turning off the people we need most to understand the situation and who must act on what they read and hear.
>
> As difficult as it is, however, for journalists to fashion a readable narrative for complex issues without depressing our readers and viewers, there is an even harder challenge—to pierce the *ideology* [italics added] that governs official policy today. One of the biggest changes in politics in my lifetime is that the *delusional is no longer marginal.* [italics added] It has come in from the fringe, to sit in the seat of power in the oval office and in Congress. For the first time in our history, ideology and theology hold a monopoly of power in Washington. Theology asserts propositions that cannot be proven true; ideologues hold stoutly to a worldview despite being contradicted by what is generally accepted as reality. When ideology and theology couple, their offspring are not always bad but they are always blind. And there is the danger: Voters and politicians alike become oblivious to the facts.
>
> Remember James Watt, President Reagan's first secretary of the Interior? My favorite online environmental journal, the ever-engaging *Grist*, reminded us recently of how James Watt told the U.S. Congress that protecting natural resources was unimportant in light of the imminent return of Jesus Christ . . . .

He is a member of the American Academy of Arts and Letters and has been the recipient of numerous honorary degrees.

41 Reprinted with permission from Harvard University, Center for Health and the Global Environment.

Beltway elites snickered. The press corps didn't know what he was talking about. But James Watt was serious. So were his compatriots out across the country. They are the people who believe the Bible is literally true-one-third of the American electorate, if a recent Gallup poll is accurate. In this past election several million good and decent citizens went to the polls believing in the rapture index. That's right—the rapture index. Google it and you will find that some of the best-selling books in America today are the 12 volumes of the Left Behind series written by the Christian fundamentalist and religious right warrior, Timothy LaHaye. These true believers subscribe to a fantastical theology concocted in the 19th century by a couple of immigrant preachers who took disparate passages from the Bible and wove them into a narrative that has captivated the imagination of millions of Americans.

Its outline is rather simple, if bizarre (the British writer George Monbiot recently did a brilliant dissection of it and I am indebted to him for adding to my own understanding): Once Israel has occupied the rest of its "biblical lands," legions of the Antichrist will attack it, triggering a final showdown in the valley of Armageddon. As the Jews who have not been converted are burned, the Messiah will return for the rapture. True believers will be lifted out of their clothes and transported to heaven, where, seated next to the right hand of God, they will watch their political and religious opponents suffer plagues of boils, sores, locusts, and frogs during the several years of tribulation that follow.

I'm not making this up. Like Monbiot, I've read the literature. I've reported on these people, following some of them from Texas to the West Bank. They are sincere, serious, and polite as they tell you they feel called to help bring the rapture on as fulfillment of biblical prophecy. That's why they have declared solidarity with Israel and the Jewish settlements and backed up their support with money and volunteers. It's why the invasion of Iraq for them was a warm-up act, predicted in the Book of Revelation where four angels "which are bound in the great river Euphrates will be released to slay the third part of man." A war with Islam in the Middle East is not something to be feared but welcomed—an essential conflagration on the road to redemption. The last time I Googled it, the rapture index stood at 144—just one point below the critical threshold when the whole thing will blow, the son of God will return, the righteous will enter heaven, and sinners will be condemned to eternal hellfire.

So what does this mean for public policy and the environment? Go to *Grist* to read a remarkable work of reporting by the journalist Glenn Scherer—"The Road to Environmental Apocalypse." Read it and you will see how millions of Christian fundamentalists may believe that environmental destruction is not only to be disregarded but actually welcomed—even hastened—as a sign of the coming apocalypse . . . . Senator Zell Miller of Georgia . . . recently quoted from the biblical book of Amos on the Senate floor: "The days will come, sayeth the Lord God, that I will send a famine in the land." He seemed to be relishing the thought.

And why not? There is a constituency for it. A 2002 *Time*/CNN poll found that *59 percent of Americans believe that the prophecies found in the Book of Revelation are going to come true.* [italics added] Nearly one-quarter think the Bible predicted the 9/11 attacks. Drive across the country with your radio tuned to the more than 1,600 Christian radio stations or in the motel turn to some of the 250 Christian TV stations and you can hear some of this end-time gospel. And you will come to understand why people under the spell of such potent prophecies cannot be expected, as *Grist* puts it, "to worry about the environment." Why care about the earth when the droughts, floods, famine, and pestilence brought by ecological collapse are signs of the apocalypse foretold in the Bible? Why care about global climate change when you and yours will be rescued in the rapture? And why care about converting from oil to solar when the same God who performed the miracle of the loaves and fishes can whip up a few billion barrels of light crude with a word?

Because these people believe that until Christ does return, the Lord will provide. One of their texts is a high-school history book, *America's Providential History.* You'll find there these words: "The secular or socialist has a limited resource mentality and views the world as a pie . . . that needs to be cut up so everyone can get a piece." However, "The Christian knows that the potential in God is unlimited and that there is no shortage of resources in God's earth . . . while many secularists view the world as overpopulated, Christians know that God has made the earth sufficiently large with plenty of resources to accommodate all of the people." No wonder Karl Rove goes around the White House whistling that militant hymn, "Onward Christian Soldiers." He turned out millions of the foot soldiers on November 2, including many who have made the apocalypse a powerful driving force in modern American politics.

I can see in the look on your faces just how hard it is for the journalist to report a story like this with any credibility. So let me put it on a personal level. I don't know how to be in this world without expecting a confident future and getting up every morning to do what I can to bring it about. So I have always been an optimist. Now, however, I think of my friend on Wall Street whom I once asked: "What do you think of the market?" "I'm optimistic," he answered. "Then why do you look so worried?" And he answered: "Because I am not sure my optimism is justified."

I'm not, either. Once upon a time I agreed with Eric Chivian and the Center for Health and the Global Environment that people will protect the natural environment when they realize its importance to their health and to the health and lives of their children. Now I am not so sure. It's not that I don't want to believe that—it's just that I read the news and connect the dots . . . .

I read the news just this week and learned how the Environmental Protection Agency had planned to spend $9 million ($2 million of it from the administration's friends at the American Chemistry Council) to pay poor families to continue to use pesticides in their homes. These pesticides have been linked to neurological damage in children, but instead of ordering an end to their use, the government and the industry were going to offer the families $970 each, as well as a camcorder and children's clothing, to serve as guinea pigs for the study . . . .

I read the news just last night and learned that the administration's friends at the International Policy Network, which is supported by Exxon Mobil and others of like mind, have issued a new report that climate change is "a myth, sea levels are not rising, scientists who *believe* [italics added] catastrophe is possible are an embarrassment."

I not only read the news but the fine print of the recent appropriations bill passed by Congress, with the obscure (and obscene) riders attached to it: a clause removing all endangered species protections from pesticides; language prohibiting judicial review for a forest in Oregon; a waiver of environmental review for grazing permits on public lands; a rider pressed by developers to weaken protection for crucial habitats in California.

I read all this and look up at the pictures on my desk, next to the computer—pictures of my grandchildren: Henry, age 12; Thomas, age

10; Nancy, 7; Jassie, 3; SaraJane, 9 months. I see the future looking back at me from those photographs and I say, "Father, forgive us, for we know not what we do." And then I am stopped short by the thought: "That's not right. We do know what we are doing. We are stealing their future. Betraying their trust. Despoiling their world."

And I ask myself: Why? Is it because we don't care? Because we are greedy? Because we have lost our capacity for outrage, our ability to sustain indignation at injustice? What has happened to our moral imagination?

On the heath Lear asks Gloucester: "How do you see the world?" And Gloucester, who is blind, answers: "I see it feelingly.'"

"I see it feelingly."

Why don't we feel the world enough to save it—for our kin to come? This brings me back to the Center, to Dr. Chivian, and to all of you gathered here this evening. You are the antidote to despair, the cure for cynicism, the answer to the faces of my grandchildren looking back at me from those pictures on my desk. Your work for the science of human health is reinforced by what the ancient Israelites called *hochma*—the science of the heart—the capacity to see, feel, and then to act as if the future depended on you.

Because it does.

—Bill Moyers

These are serious indictments being made by well-meaning people in the know who normally reserve judgment, but they, like me, are observing a trend that will eventually catapult us back into the Dark Ages. Even as I write this in April of 2009, Congress attempted to put a bill (HR1388) on the floor for a vote that would have required mandatory military and/or community service for our youth. There was so much public anger over the bill that they needed to remove "mandatory service" to get the bill passed. What did they do behind the scenes? They wrote another bill (HR1444) that includes "mandatory service." Why is it that our government is continually attempting to pass bills such as this against the will of the people? The answer is complex but the main idea is to blur the line between the military and civilian aspects of our society. It would also allow for putting civilians on our streets with a military agenda that could operate outside the restrictions of the Constitution. Do your own research and Google "HR1388 and HR1444."

A question to Albert Einstein after he witnessed what the atomic bomb could do was: *"What do you think we will be fighting World War III with?"* His answer was, *"I don't know, but I know what we will be fighting World War IV with—sticks and stones."*

If this is your estimate of your relation to nature and you have an advanced technology, your likelihood of survival will be that of a snowball in hell. You will die either of the toxic by-products of your own hate, or, simply, of overpopulation and overgrazing.
Gregory Bateson, futurist and naturalist

CHAPTER 14

Where Are the Young Scientists in America? A Talk with Fred Alan Wolf (Dr. Quantum)

One of the concerns that comes across loud and clear with people such as Bill Moyers is children. The question of children also came up when I was talking with Fred Alan Wolf, who wrote *Parallel Universes.*[42] Having the opportunity to read books that expand a child's awareness is imperative in today's world. The problem is that they are not getting this type of information in public schools and this lack of information appears to be part of the indoctrination process we see going on today. Without the proper input a person simply doesn't develop the desire to know more. Without the desire to know more you will forever maintain old dysfunctional beliefs. This is a small part of my conversation with one of today's leading physicists and teachers.

---

[42] Fred Alan Wolf is a physicist, writer, and lecturer who earned his PhD in theoretical physics at UCLA in 1963. He continues to write, lecture throughout the world, and conduct research on the relationship of quantum physics to consciousness. He is the National Book Award-winning author of *Taking the Quantum Leap* (1981, 1989). He's also the author of many other books, including *Parallel Universes* (1990), *The Dreaming Universe* (1995), *The Eagle's Quest* (1992), *The Spiritual Universe* (1998), *Mind into Matter: A New Alchemy of Science and Spirit* (2000), *Matter into Feeling: A New Alchemy of Science and Spirit (2002)*, *The Yoga of Time Travel: How the Mind Can Defeat Time* (2004), and his latest book, *Dr. Quantum Presents: A Little Book of Big Ideas* (2005). For more information visit www.fredalanwolf.com.

Rahasya: I was reading that the United States is producing fewer young scientists in recent years than it ever has and I couldn't help but think that the Dr. Quantum series you publish could help spark young minds.

Fred Alan Wolf: I truly think it should, and that's my greatest hope, to stir minds both young and old. One of my main reasons for doing this is to influence young minds. I have done so in the past, even back when my first book came out, called *Taking the Quantum Leap*. That influenced a lot of young minds. Kids who were in high school then are now physicists and they've told me that they were very influenced by *Taking the Quantum Leap*. So, the Dr. Quantum series should help along those lines also.

Rahasya: Yes. It influenced me. That was one of the first books I read that led me into the metaphysical realms.

Fred Alan Wolf: Reading allows you to enter into another realm, another mind, and see life differently through the eyes of someone else.

Rahasya: When we brought the movie* What the Bleep Do We Know? *to our community, it played for seven weeks and sold out most of that time, so we are ready and open for a paradigm shift of how we see this world. That being said, I also know how deeply ingrained old patterns are and how easy it is to fall into them. What would you suggest that the average person could do to break down old beliefs and open up to a new worldview?

Fred Alan Wolf: The first thing we need to do is realize that some of our old beliefs are dysfunctional and no longer work, if indeed they ever did. If the desire to do that isn't there, there's nothing to do. If you don't want to do something, you're not going to do it. So there's no way to take an average person to this experience of awakening without the person having desire first. However, if you have that desire, then there are all kinds of ways to pursue your awakening. And there are many different paths, whether it's a spiritual path, or a path through science, or a path through yoga, there are many ways of getting there. So my first thought would be, if you haven't awakened that desire in yourself, then start hanging out with people who have awakened that desire in themselves, and see if that will be contagious enough to get you to get off your butt and do something about it.

Rahasya: That's great advice. One of the first things I noticed in my life was that most of my friends and acquaintances faded away when I decided to take a different path and I now find it very important to be around people who . . . don't drag me down.

Fred Alan Wolf: And they are probably still there waiting for you should you ever decide to go backwards to a less spiritual way of living. The main thing is you gotta take action. You can't sit there and hope it's going to happen.

Rahasya: Yes, I think that's the importance of living in communities that are progressive in their thinking. I believe one of the reasons for my magazine's success has been the people living in our Northern California regions.

Fred Alan Wolf: Yes, and it's important to have a free press where you can express ideas that may not be mainstream or popular.

Rahasya: I think that's one of the drawbacks of mainstream media. They're so dependent on their advertisers and most of their ads promote pharmaceuticals, processed foods, alcohol, and the list goes on.

Fred Alan Wolf: That's an important point in today's world so keep it up.

One of the things I remember about my conversation with Fred Alan Wolf is a feeling of urgency to prepare our children for the world they are growing up in, but we're not doing that for a number of reasons. One of the main financial reasons at this point is that we have spent well over $1 trillion on the war in the Middle East and some say it will cost at least that much more. This is money that we should be spending quite differently but we are again caught in the social consequences of different belief systems' colliding with each other and culminating in war as they always do.

When I think of programs such as No Child Left Behind, it makes me sick to think so many Americans buy into this political propaganda that's meant to sedate us and keep us sleeping in the "American Dream." A movie to see on this subject is *Teacher,* starring Nick Nolte and produced by Aaron Russo, who also produced *From Freedom to Fascism.* At the time of this writing only 80 percent of white children graduate from high school, 60 percent of African Americans, and 62 percent of Hispanics. This is absolutely embarrassing as an American yet most of the population has its attention drawn to what's going on in other countries. What compounds this problem is the fact that the information we get is filtered and spun to match whatever the political agenda of the day happens to be. I remember reading that Oprah Winfrey was looking at schools in the United States to see how she could best spend millions of her dollars and she received a lot of negative feedback when she decided to go to Africa and build a school there. She said when she went around to schools here and asked the students what they wanted most and how she could help, one of the priorities was they wanted iPods. When she went to Africa they wanted clothes, shoes, books, and food. So the question is, "What are we teaching our children?" The simple answer to that is we are not teaching them anything meaningful. They are learning from the mass media that thrives on consumerism. This is why the average grade-school child can name hundreds of company logos but fails to name the three branches of government.

CHAPTER 15

What Do Religious Moderates Have to Do with This?

Now—I can already hear you saying, "But most religious people are moderates and don't directly cause these problems we see." But I say that standing by with apathy and quietly supporting these religious movements and political agendas by your silence is probably the biggest problem of all. It was the Catholic Church that stood by and watched while Hitler came to power and positioned himself to do what he did. To this day the Catholic Church has never excommunicated one person involved in the Holocaust. This isn't to say that there weren't many brave and courageous Christians throughout history doing heroic deeds, but that alone does not prove a religious doctrine is true. Good Southern Christians stood by and watched and even took part in the Ku Klux Klan and its agendas while moderates stood by, silently watching and sometimes inwardly applauding its actions through their own inaction. I know this is difficult to look at and I know how easy it is to "cherry-pick" all the good things religious people do to support your point of view, but this approach is valid only for those attempting to prove themselves right, but we are looking for the truth no matter where it leads.

The price of apathy toward public affairs is to be ruled by evil men.
Plato (428-448 BC)

Don't think for a moment that most Christians, Jews, and Muslims don't harbor a dark secret as far the idea of "others" goes. Ask most moderates (I have asked quite a few) what they would do if they had a magic wand that would somehow transform the "others" into their own religion or remove them from the planet for God to judge? Most say they would wave that wand from every corner of the planet until there were no "others" left. So it's not a matter of *what* they would do; it's a matter of *how much suffering* they are willing to cause to do it. And how much suffering a person is willing to cause is usually dependent on how much that person is actually aware of on a personal level. Most of us know that well over 25,000 children died of starvation today but that would not affect us nearly as much as our neighbor's child dying of starvation. This is a dangerous situation in a technologically advanced world where most wars are decided on and fought without our direct involvement and with the push of a button or a vote in Congress for more funding. How many of us watched the "shock and awe" attack on Baghdad between advertisements selling us SUVs that get 12 miles to a gallon or some pharmaceutical company selling us a drug for stress? As individuals we now hide behind our technology the same way we have always hidden in crowds and mobs. Of course we have evolved and call them organizations, corporations, nations, and churches. This is why when we watch our news programs they don't allow us to see the carnage, especially that of our own soldiers and of the innocent people who are killed. Please, don't let these points slide by without really understanding their deeper implications. It's not only what we do and don't do that make a difference; it's also those deep-seated beliefs of how we see ourselves and others in this world that also make a difference.

Whatever you do is going to be seen as evil to somebody.
Joseph Campbell (1904-1987)

It's worth noting that most of the proposed solutions of the day sound good on the surface, but if you look a little deeper you will see that they usually have to do with converting the rest of the world to one's personal beliefs. It is also worth noting that if you have even the slightest idea that it would be possible to convert the whole world to your point of view, you are living in a self-imposed delusion of massive proportions. But don't feel alone, because if you look at the actions of most religions, you will see that they not only act as if they can but it is built into their dogma that they must convert or destroy nonbelievers and infidels or die in glory trying.[43] And this is all done with the belief that they are pleasing their loving God.

---

[43] Deuteronomy 13: 1: If there arise among you a prophet, or a dreamer of dreams, and giveth thee a sign or a wonder, 2: And the sign or the wonder come to pass,

If we're going to be absolutely truthful with ourselves we have to admit that there's not a single person on this planet who has good reason to believe that Jesus rose from the dead or that the angel Gabriel came to Muhammad and spoke to him in a cave.[44] Yet there are billions of people worldwide who organize their entire lives around these beliefs. As a direct result we live in a world divided by concepts and beliefs about sex, eternal life, the value of being a man or a woman, and many other ideas, including ideas about when

---

whereof he spake unto thee, saying, Let us go after other gods, which thou hast not known, and let us serve them; 3: Thou shalt not hearken unto the words of that prophet, or that dreamer of dreams: for the LORD your God proveth you, to know whether ye love the LORD your God with all your heart and with all your soul. 4: Ye shall walk after the LORD your God, *and fear him*, and keep his commandments, and obey his voice, and ye shall serve him, and cleave unto him. 5: And that prophet, or that dreamer of dreams, *shall be put to death;* because he hath spoken to turn you away from the LORD your God, which brought you out of the land of Egypt, and redeemed you out of the house of bondage, to thrust thee out of the way which the LORD thy God commanded thee to walk in. So shalt thou put the evil away from the midst of thee. 6: If thy *brother, the son of thy mother,* or thy *son,* or thy *daughter,* or the *wife* of thy bosom, or thy *friend,* which is as thine own soul, entice thee secretly, saying, Let us go and serve other gods, which thou hast not known, thou, nor thy fathers; 7: Namely, of the gods of the people which are round about you, nigh unto thee, or far off from thee, from the one end of the earth even unto the other end of the earth; 8: Thou shalt not consent unto him, nor hearken unto him; neither shall thine eye pity him, neither shalt thou spare, neither shalt thou conceal him: 9: But *thou shalt surely kill him;* thine hand shall be first upon him to put him to death, and *afterwards the hand of all the people.* 10: And *thou shalt stone him with stones,* that he *die;* because he hath sought to thrust thee away from the LORD thy God, which brought thee out of the land of Egypt, from the house of bondage.

44 How could there be empirical evidence that Muhammad "didn't" leave the planet on a winged horse or was spoken to by an angel? This is where we need to make religions be accountable because it's not up to someone like me to prove that a person 2,000 years ago didn't rise from the dead when in fact no one ever has that can be validated. If I say there is a teapot orbiting Jupiter it's not up to everyone else to disprove it; it's up to me to prove it. In rational circles this concept is called Russell's Teapot, a name coined by Bertrand Russell.

the world will end, which as mentioned before has caused untold damage to our environment.

Within every religion are different levels of commitment. Within the inner circle you find Muslim suicide bombers waiting in line to kill innocent civilians and Dominionist Christians who have no problem calling for the execution of homosexuals, foreign presidents, and blasphemers. Beyond the inner circle you will find millions of followers who share the views of the inner circle but don't have the commitment it takes to blow themselves up or put a noose around a person's neck for using God's name in vain. Further out still are the moderates who have cherry-picked their scriptures to pieces to live in the modern world and not appear to be too out of touch with reality or get arrested by the local authorities or placed in mental institutions.

Now we get to the liberals, who at first glance seem to be thinking for themselves. The problem with all these different levels of people's commitment to their beliefs is that they all support the inner circle of fanatics who are making every effort to turn our world into a totalitarian theocracy. Even though moderates and liberals will often admit the insanity of the beliefs of their more rigid brethren, they are not willing or have the courage to step up and speak out openly. I can't begin to tell you the number of people I have interviewed and talked with who have told me things they believe and don't believe but who have asked me to not make their beliefs public. We are still living in the shadow of persecution from religious fundamentalism and even today that shadow becomes all too real to people such as Benazir Bhutto, who was assassinated for believing in the equality of women and democracy and speaking out against the Taliban and Pakistan, where she was twice prime minister (1988-1990 and 1993-1996).[45]

The End Times

One aspect of the Judeo-Christian religion is often overlooked. It's called "dispensationalism," for those of you who want to look it up and go deeper. Basically it's a way of reading the Bible in which there are two stories. There's the story of God's earthly people, the Israelites, and there's

---

[45] Ms. Bhutto, 54, was shot in the neck or head, according to differing accounts, as she stood in the open sunroof of a car and waved to crowds. Seconds later a suicide attacker detonated his bomb, killing more than 20 people and wounding 50.

the story of God's heavenly people, the Church. This is why the prophecies concerning earthly events apply to the Jewish people. Almost all prophecies lead to and are fulfilled in the "end times." One of the prophecies is that the Jewish people must return to their little piece of real estate that God gave them and they must stay there. This needs to happen before there can be an Antichrist, the destruction and rebuilding of the Temple Mount in Jerusalem, the final battle of Armageddon, where the blood will be a couple of feet deep, and of course the return of Jesus Christ, who will take revenge on the rest of us who don't happen to believe this by sending us to the pits of hell to suffer for eternity. So the way it is now, the Christians with Christian armies have a sacred duty to reinstall the Jews in their Holy Land so Jesus Christ, their savior, can come back, swoop them all up, and take them off to heaven, where they can watch the horrible consequences happening to all those who didn't "hear the word." Where this story gets a little deceptive in regard to the Jewish people, and which shouldn't be overlooked, is that there's a catch to all this help from the Christians; when Jesus comes back, all the Jews get cast down into the fiery depths of hell for eternal suffering unless they accept Jesus as their savior. And of course the Muslims are pretty much . . . well, let's just say they're in a bad situation no matter what, unless of course Muhammad comes back first on his winged horse. So when Jesus comes back it may be with a sword but it will have a double edge for most of us. This is one of the many examples of what happens when we apply reason and logic to a belief and extrapolate it out to its final objective; it's almost always without exception ending up in some absurd, impossible, and inhospitable place with repulsive consequences.

I don't believe Jesus came to start Christianity.
Oprah Winfrey, talking to Eckhart Tolle

Another aspect of this "end-time" situation is that you need to remember that the Jews are still waiting for their messiah and the Christians are waiting for the Antichrist to emerge, both of whom are said to have some of the same powers to bring peace to this planet. What kind of madness could this evoke in the world? And what if society simply produces a great man with great powers of love, compassion, and intelligence who is trying to help the world by unifying nations, as most New Agers believe? Some of us will admire him; some will try to kill him as the Antichrist, and some will claim him as a savior and the messiah they've been waiting for. So if you claim to be a religious moderate, tell me, "How do you reconcile any of this into any kind of moderation?" In the whole recorded history of mankind, we have never been able to moderate madness and insanity, only contain it so it doesn't gain control of our lives.

Once started, religious strife has a tendency to go on and on, to become permanent feuds. Today we see such intractable interreligious wars in Northern Ireland, between Jews and Muslims and Christians in Palestine, Hindus and Muslims in South Asia and in many other places. Attempts to bring about peace have failed again and again. Always the extremist elements invoking past injustices, imagined or real, will succeed in torpedoing the peace efforts and bringing about another bout of hostility.
Datuk Seri Dr Mahathir Mohamad, prime minister of Malaysia, addressing the World Evangelical Fellowship on May 4, 2001

It's also worth mentioning to religious moderates that the extremists in your religion and other religions really do believe the things they say they believe. What I see as lunacy and madness, and you as a moderate don't see at all or discount as unimportant, they see as gospel (this means *God's spell,* by the way). I bring this up because I found myself writing about the Catholic belief that a priest can actually turn a small cracker into the flesh of a man born 2,000 years ago. I simply assumed that the priests really didn't actually believe it until I watched a documentary on what has been going on in the priesthood with alter boys; I listened to the priests themselves say that this is an absolute fact and beyond belief because they know the wafer turns into the flesh of Jesus Christ. They say the ability they have to do this—the ability to perform the Eucharist—is what separates them from normal human beings.[46] It hit me then that there was a part of me that didn't

[46] For those of you who need a little validation as to what the Catholic Church says, here is what the Council of Trent says about the Eucharist: "I likewise profess that in the Mass a true, proper and propitiatory sacrifice is offered to God on behalf of the living and the dead, and that the body and blood together with the soul and divinity of our Lord Jesus Christ is truly, really, and substantially present in the most holy sacrament of the Eucharist, and that there is a change of the whole substance of the bread into the body, and of the whole substance of the wine into blood; and this change the Catholic Church calls transubstantiation. I also profess that the whole and entire Christ and a

fully understand or realize people could actually believe this . . . but they do. My point is this: Don't think for a moment that Muslim fundamental extremists don't really believe that they will have 72 virgins and spend an eternity having orgasms in paradise if they strap on a bomb and blow up infidels. It's a sobering thought to consider that hundreds of millions of people on this planet see my death as their way to eternal orgasmic bliss. And don't think for a second that there are not several million American Christians who see the dropping of an atomic bomb in the Middle East as a good thing because it means they are getting closer to the time when Jesus comes down out of the clouds to whisk them away to paradise and send the rest of us nonbelievers to eternal damnation (several polls throughout the United States confirm this). Of course the Muslims are fully expecting Muhammad to return on the winged horse that he left on.

Such evil deeds could religion prompt.
Lucretius (96 BC-55 BC), *De Rerum Natura*

It's also interesting to see that all three world religions can no longer live in a modern society by the letter of the law in their Holy Books because so many of their laws and commandments are now illegal and indeed, immoral, by most of today's standards. This alone is very interesting to realize, that we, as a society, have evolved morally past our religious dogmas in the face of persecution, excommunication, and even torture and death. *Religious zealots today would find themselves in jail, a mental hospital, or Guantanamo Bay detention camp if they attempted to practice the old laws found in their Holy Books.* Imagine your neighbor's dropping by and asking for a knife because he heard the voice of God telling him to kill his son to prove his obedience.[47] The same people we used to make into our prophets would be locked up today for performing the very same acts that once proved their divine connection with a very capricious, vengeful, jealous, and strange God, the same God they call the "God of love." By definition, many of the prophets in the old Bible would be considered mass murderers and be convicted for

---

true sacrament is received under each separate species." For a very in-depth look into the original meaning of this sacrament, read *Bloodline of the Holy Grail* (1997) and *Genesis of the Grail Kings* (2002) by Laurence Gardner.

47 Genesis 22: And Abraham stretched forth his hand, and took the knife to slay his son.

war crimes in today's world. Society is straining to evolve past these types of beliefs. History has shown what happens when religion takes over and reason is put aside; this is the period we now call the Dark Ages.

I think we ought always to entertain our opinions with some measure of doubt. I shouldn't wish people dogmatically to believe any philosophy, not even mine.
Bertrand Russell (1872-1970)

I have mentioned that I have no intention to be antagonistic, aggressive, or to convince you of something without proof so if you feel attacked, at least consider the possibility that this feeling is coming from your own fears and your own attempt to defend your point of view. Keep in mind that you can defend only something that came about through some rational process. I made quite a few attempts to write in such a way that I wouldn't offend too many people but quickly realized there was no way to do that, because a lot of people on this planet have put themselves in a situation that is quite offensive even to themselves. In reality there is simply no way to write about something that is irrational and absurd without seeming antagonistic and condescending. Besides, it's possible that we are running out of time and no longer have the option or luxury of playing out our nightmares as if there will be a constant replenishing of our tomorrows.

With or without religion, you would have good people doing good things and evil people doing evil things. But for good people to do evil things, that takes religion.
Steven Weinberg

CHAPTER 16

A Special Note to Christians Living in the United States

I see our country on a dangerous path that could leave us in a spiritual darkness. You may be thinking that by holding on to traditional "old-time religion" you are helping to hold the light and ward off the darkness and the evildoers. But the trouble is that with all the advancement in science and human understanding, we are undermining much of the logic and rationale for what religion says is true. In fact, most of it simply can't be true according to our more recent knowledge and experience. What I see happening was put best by Joseph Campbell when he said his concern was that as scientific advancement progresses we will have the tendency to do away with religion altogether. This means that we will also be doing away with all the symbolic and mythological knowledge embedded in the stories, which history has shown are the best transmitter of deeper understanding and esoteric truths through multiple generations and different cultures. If this happens, we will end up with a society with no mythology, and most societies are held together through a common myth or story. So if we don't take action in updating our concepts of religion, we risk the danger of losing it all, which would be the beginning of our unraveling as a society. Timothy Freke calls this "throwing out the baby with the bathwater" in his book, *The Laughing Jesus* (2005). We will hear more from him later.

"We are developing so quickly on this planet that we are not developing new myths. The future myth will be the planet itself."
Joseph Campbell (1904-1987)

Of course we wouldn't be the first nation to collapse because of holding on to outdated ideas and beliefs. It actually happens regularly throughout history and we would be an exception to the rule if it does not happen to us. So it may be a matter of learning from history so we need not repeat it. Almost every culture throughout history collapsed because it lost touch with the advancement of the world around it. There are other predictors that we are also embracing as a nation, such as our imperialistic desire to expand our culture and beliefs throughout the world, our depletion of natural resources within our national boundaries, and a national debt that can never be repaid. The mere fact that our national debt is a creation of an illegal Federal Reserve created for the sole purpose of financially enslaving this and other nations should do away with our insane preoccupation with nationalistic beliefs, but it doesn't because people believe what they are told to believe.[48] Now we are on the verge of financial collapse, with China buying up our credit-card companies, real estate, and other investments while we buy toxic toys and poison food from it. We even have other countries buying our infrastructures, such as turnpikes and historical buildings.

We have, as they say, "taken it to the wire." Our whole planet is on the brink of social, financial, and ecological collapse, and to a large degree it has to do with our political, religious, and personal beliefs. We hear every day about mad suicide bombers who are trying to blow up our children to the point we don't even listen any more. And please don't tell me it's because of how they (the suicide bombers) were mistreated, or that they are going hungry and this is their only alternative. If this were a good excuse, consider that the Tibetan people went through much more suffering and hardship and I have never heard of a Buddhist suicide bomber and never will, and this is because they do not "believe" they are commanded by a God to blow themselves up and that they will go to a better place by doing so; if they did, be assured that we would be in even more trouble than we are today.

And just in case you think you have the corner on saviors with all the attributes that Jesus had, think again. A multitude of "sons of Gods" from every continent and every culture share many of the same mythological

---

[48] A good resource here is *From Freedom to Fascism* (2006), a documentary by Aaron Russo. You can buy it at www.freedomtofascism.com or watch it free on You Tube. Also go to www.zeitgeistmovie.com and watch the first video; the second, *Addendum* (2008), is about national and global financial institutions.

attributes, Horus, for example. I will list his similarities to Jesus but I could list many other "Gods" throughout history who share these same attributes.

Jesus/Horus

- Jesus was conceived of a virgin. (Matthew 1:23 and Luke 1:27)
- Horus was conceived of a virgin.

- Jesus was the "only begotten son" of the god Yahweh. (Mark 1:11)
- Horus was the "only begotten son" of the god Osiris.

- The foster father of Jesus was Joseph.
- The foster father of Horus was Jo-Seph.

- Joseph was of royal descent, being from the House of David.
- Jo-Seph was of royal descent.

- The coming birth of Jesus was announced to Mary by an angel. (Luke 1:34)
- The coming birth of Horus was announced to his mother by an angel.

- The birth of Jesus was heralded by a star in the East (where the sun rises in the morning). (Matthew 2:2 and Matthew 2:9)
- The birth of Horus was heralded by the star Sirius (the morning star).

- The birth of Jesus was announced by an angel. (Luke 2)
- The birth of Horus was announced by an angel.

- Jesus was visited by shepherds at his birth. (Luke 2)
- Horus was visited by shepherds at his birth.

- Jesus was visited by magi (astrologers or wise men) at his birth. Tradition says there were three of them. (Matthew 2)
- Horus was visited by "three solar deities" at his birth.

- Jesus has no official recorded history between ages 12 and 30.
- Horus has no official recorded history between ages 12 and 30.

- Jesus was baptized at age 30.
- Horus was baptized at age 30.

- Jesus was baptized by John the Baptist. (Matthew, Mark, and Luke)
- Horus was baptized by Anup the Baptizer.

- Jesus was taken from the desert in Palestine up a high mountain to be tempted by his archnemesis Satan. (Mark 1:13)
- Horus was taken from the desert of Amenta up a high mountain to be tempted by his archrival Set.

- Jesus successfully resists.
- Horus successfully resists.

- Jesus has 12 disciples, although their names are in dispute.
- Horus has 12 disciples.

- Jesus walked on water (Matthew 14:22, Mark 6:45, John 6:16), cast out demons (Mark 7:26), healed the sick (Matthew 4:23, Mark 1:32), and restored sight to the blind (John 9).
- Horus walked on water, cast out demons, healed the sick, and restored sight to the blind.

- Lazarus was raised, or at least lived, in Bethany (literally, "house of Anu"). (John 12:1)
- Osiris was raised in the town of Anu.

- Jesus delivered a Sermon on the Mount.
- Horus delivered a Sermon on the Mount.

- Jesus was crucified.
- Horus was crucified.

- Jesus was crucified next to two thieves.
- Horus was crucified next to two thieves.

- Jesus was buried in a tomb. (John 19:42)
- Horus was buried in a tomb.

- Jesus was known as the Christ (which means "anointed one").
- Horus is known as KRST, the anointed one.

- Jesus has been called the good shepherd (John 10:11, 10:14), the lamb of God (John 1:29, 1:36), the bread of life (John 6:35, John 6:48), the son of man (many places), the Word, the fisher, and the winnower.
- Horus has been called the good shepherd, the lamb of God, the bread of life, the son of man, the Word, the fisher, and the winnower.

- Jesus is associated with the zodiac sign of Pisces (the fish).
- Horus is associated with the zodiac sign of Pisces (the fish).

- Jesus was born in Bethlehem ("the house of bread"). (Matthew 2:1)
- Horus was born in Anu ("the place of bread").

- Jesus was transfigured on the mount.
- Horus was transfigured on the mount.

- Jesus is identified with the Tau (cross).
- Horus was identified with the Tau (cross).

As I mentioned before, the similarities are also numerous with many other Biblical figures, especially Moses, but I believe this list makes the point, which is that there is a common and quite popular mythology based on the movement of the planets and the changing seasons.

CHAPTER 17

Women of the World—Wake Up: It's Your Moment!

If anything is going to be the undoing of religious beliefs and political ideologies it will be the women of the world standing up for their rights as equal human beings. Since we live in a world that is predominately ruled by men, we also will need men to stand up and be counted. Women have suffered unspeakable injustice, cruelty, genital mutilation, honor killing, polygamy, rape and even punishment for being raped, forced marriages, burnings at the stake, and up until recent times, even in the United States, they were not allowed to hold office or vote or even to have a bank account in their own names, and the list goes on.

At the time of this writing a woman in Saudi Arabia was raped by seven men while out by herself. What do you think happened? Keep in mind that we live in the 21st century. She was sentenced to prison by the judge and to be given 90 lashes. When her lawyer complained that the punishment was unjust the judge changed the punishment to 200 lashes. But what about that woman? What about the emotional scars that will last a lifetime? And here is the strange and sad part of the story . . . she is one of the lucky ones; she has a husband who sees the unjustness and insanity and is trying to get her released. Now I hear you asking, "What was the justification behind this decision?" It was the fact that the woman was out by herself. In the Muslim religion a woman can not be out by herself, or admit herself into a hospital without a man present to sign her in, or even to talk to a neighbor unless it's absolutely necessary. Of course, especially in Western societies, some of these rules are, out of moral necessity, being relaxed somewhat but the point is that these rules are still being used to abuse and in some cases kill women.

How did this start? Keep in mind that the Judeo-Christian-Islamic religions all have the same basic teachings as far as women go, although the Muslims' attitude is more extreme and they actually carry out what their Holy Books say. First of all, it was man who was made in the image of God; women were little more than an afterthought and women were made from man so they are one step removed from God (Genesis 2:21-22; in the Quran 4:1, 39:6, 7:189), much like a photocopy that is never quite as clear as the original. Of course a woman was also the first to bring sin into the world (Genesis 3:12). The Old Testament even sets a monetary value for a woman at one-half to two-thirds that of a man (Leviticus 27). The Quran requires two women's testimony to equal one man's testimony (2:282) and every female only receives one-half her brother's share of an inheritance (4:11).

Women have been looked upon as nothing more than a material possession, along with a man's oxen, house, and his ass, as stated in the 10th commandment.[49] The Quran even dictates that a disobedient wife should be whipped (4:34). St. Paul makes it very plain what a wife is when he states: *"Wives, be subject to your husbands, as to the Lord. For the husband is the head of the wife as Christ is the head of the church, his body, and is himself its Savior. As the church is subject to Christ, so let wives also be subject in everything to their husbands."* (Ephesians 5:22-24). If this isn't a classic example of a purely man-made rule, I don't know what is.

One of the most influential Muslims since Muhammad was Al-Ghazali (1058-1111), who lays out precisely what is expected of a woman when he writes:

> *She should stay home and get on with her spinning, she should not go out often, she must not be well-informed, nor must she be communicative with her neighbors and only visit them when absolutely necessary; she should take care of her husband and respect him in his presence and his absence and seek to satisfy him in everything . . . she must not leave the house without his permission and if given his permission she must leave surreptitiously. She should put on old clothes and take deserted streets and alleys, avoid markets, and make sure that a stranger does not hear her voice or recognize her;*

---

49 The 10th Commandment (Exodus 20:17): Thou shalt not covet thy neighbor's house, thou shalt not covet thy neighbor's wife, nor his manservant, nor his maidservant, nor his ox, nor his ass, nor any thing that [is] thy neighbor's.

she must not speak to a friend of her husband even in need . . . . Her sole worry should be her virtue, her home as well as her prayers and her fast. If a friend of her husband calls when the latter is absent she must not open the door nor reply to him in order to safeguard her and her husband's honor. She should accept what her husband gives her as sufficient sexual needs at any moment . . . . She should be clean and ready to satisfy her husband's sexual needs at any moment.
(Cited in Ibn Warraq's *Why I Am Not Muslim*, 1995, p. 300).

It's easy to see the effects of this type of belief when you look at Muslim women wearing their veils. You might be surprised to learn that this tradition came from a sect of Christians called the Ecclesiastics. And it's not just in the Muslim tradition that we find absurdities that would not be tolerated by any modern-thinking woman. We can read in the Old Testament that if a woman is raped and fails to scream loudly enough that she should be stoned to death (Deuteronomy 22:24). It's even said to be perfectly okay to slay your wife on her father's doorstep if you find out that she is not a virgin on your wedding night! But if she has proof of her virginity and her parents can show it to be true then the husband will be punished.[50] Or how about this—it's perfectly acceptable to sell your daughter into slavery.[51] Maybe I shouldn't mention this, but it's also written that if you are a Christian and

[50] Deuteronomy 22: 16: The girl's father will say to the elders, "I gave my daughter in marriage to this man, but he dislikes her. 17: Now he has slandered her and said, 'I did not find your daughter to be a virgin.' But here is the proof of my daughter's virginity. Then her parents shall display the cloth before the elders of the town, 18: and the elders shall take the man and punish him.

(The proof of virginity was often a bloody rag that could be offered as proof. This was accomplished in some cases by a thoughtful nonvirgin bride by cutting her hand and secretly bleeding on it in the process of intercourse).

[51] Exodus 12:7: If a man sells his daughter as a servant, she is not to go free as menservants do. 8: If she does not please the master who has selected her for himself, he must let her be redeemed. He has no right to sell her to foreigners, because he has broken faith with her. 9: If he selects her for his son, he must grant her the rights of a daughter. 10: If he marries another woman, he must not deprive the first one of her food, clothing, and marital rights. 11: If he does not provide her with these three things, she is to go free, without any payment of money.

someone like me tries to steer you away from your belief, then you are to slay me; as a matter of fact, if you don't, you too are in mortal danger if another Christian finds out that you didn't execute your Christian duty.

Obviously, we could go on and on with what it says in our Holy Books inspired by a loving God, but imagine for a moment how this has affected billions of women through the centuries with suffering and torment. If you are a man, imagine what it would be like if your neighbors came over and took you away to be burned at the stake for treating your family with herbs. And don't think for a moment that injustices are not happening today right here in the good old United States of America. Just recently in the news was the case of Warren Jeffs, who was found guilty in court of unlawful flight to avoid prosecution; sexual conduct with a minor; conspiracy to commit sexual conduct with a minor, including sexual conduct with minors and incest; and an accomplice to rape. But this behavior is perfectly logical if you believe what Joseph Smith wrote in *Doctrine and Covenants of the Church of Jesus Christ of Latter-Day Saints*: *"If any man espouse a virgin, and desire to espouse another, and the first give her consent; and if he espouse the second, and they are virgins, and have vowed to no other man, then he is justified."* And of course Joseph had his fair share of wives. Without making judgments for or against the tenets of this, we need to understand that this statement obviously is used to subjugate young girls into the bedroom for sex against their wills more than anyone wants to admit.

Even though the Fundamentalist Church of Jesus Christ of Latter-Day Saints is a splinter group of the Mormon religion, its members are still working under the same Holy Book and interpreting it in the same ways. The truth is that there would probably have been no splintering off if it weren't for the fact that the Mormons wanted statehood and were forced to abandon the practice of polygamy, at least out in the open. According to different polls between 25,000 to 100,000 women are living in a polygamy arrangement.

And This from Amnesty International:

Now please keep in mind that most of what follows is made possible by political regimes that prey on the religious beliefs and superstitions of their populations. Associated with almost any atrocity, such as the ones you will be reading about in this Amnesty International Report, is a religious belief that is absurd and that makes the atrocity possible by sanctifying it and sometimes demanding it.

Rape as a Tool of War: A Fact Sheet

In every armed conflict investigated by Amnesty International in 1999 and 2000, the torture of women was reported, most often in the form of sexual violence. Rape, when used as a weapon of war, is systematically employed for a variety of purposes, including intimidation, humiliation, political terror, extracting information, rewarding soldiers, and "ethnic cleansing."

Violence against women in armed conflict situations is largely based on traditional views of women as property, and often as sexual objects *as written out in Holy Books* [italics added]. Around the world, women have long been attributed the role of transmitters of culture and symbols of nation or community. Violence directed against women is often considered an attack against the values or "honor" of a society and therefore a particularly potent tool of war. Women therefore experience armed conflicts as sexual objects, as presumed emblems of national and ethnic identity, and as female members of ethnic, racial, religious, or national groups.

The consequences for victims of sexual violence in war are grave and may affect women for the rest of their lives. These include serious and chronic medical problems, psychological damage, life-threatening diseases such as HIV/AIDS, forced pregnancy, infertility, stigmatization and/or rejection by family members and communities.

Impunity for Perpetrators of Rape and Sexual Violence in War

All too often, those responsible for acts of sexual violence and rape committed in war go unpunished. Factors contributing to impunity with regard to sexual crimes in war are many, and include:

- An overall climate of indifference towards many forms of violence against women;
- The tacit acceptance of rape and other forms of sexual violence as an unavoidable part of war;
- Threats and reprisals against those who reveal abuses;
- The existence of special national legislation in many countries which prevents prosecutions for crimes committed in war;
- Laws granting amnesty to perpetrators as part of peace-making "deals."

Underreporting is also a significant barrier to justice. Many women feel shame and fear rejection from their husbands, families, and communities if they report having been raped. The threat of divorce or the possibility of being considered "unmarriable" causes many women's reluctance to report their experiences. The economic and social dependence of women on men in many societies contributes to their fear of reporting rape.

Rape in War: Specific Cases

- Since the humanitarian crisis began in 2003, women in the western Sudanese state of Darfur have been subjected to rape and other forms of gender-based violence perpetrated by the government-backed Janjawid militia, as well as other armed troops. In many cases, women have been publicly raped in front of their husbands, relatives or the wider community. Pregnant women have not been spared and those who have resisted rapes were reportedly beaten, stabbed or killed. Women and girls as young as eight years old have been abducted during attacks and forced into sexual slavery in the Janjawid military camps. The strong cultural, social, and religious taboos against rape in Darfur make women reluctant to speak out and often cause them and their children to be ostracized by their community. Raped women who are not able to marry or who have been abandoned are deprived of the "protection" and economic support that men are traditionally expected to provide in Sudan.
- In northern Uganda, the *Lord's Resistance Army* (LRA) [italics added] abducts children, forcing girls into "marriage" and institutionalized rape. Men are "given" women and girls as rewards for "good behavior," for example, following orders to kill prisoners of war and captured villagers. The men then have total sexual control over their "wives" and "domestic helpers," subjecting them to rape and various other forms of violence.
- Abduction, rape, and sexual slavery are also systematic and widespread in the conflict in Sierra Leone. Rape victims often suffer extreme brutality. In one case, a 14-year-old girl was stabbed in the vagina with a knife because she refused to have sex with the rebel combatant who abducted her. In another, a

> 16-year-old girl was so badly injured, that after her escape, she required a hysterectomy.

> The use of rape in conflict reflects the inequalities women face in their everyday lives in peacetime. Until governments take responsibility for their obligations to ensure equality, and end discrimination against women, rape will continue to be a favored weapon of the aggressor.

Of course true and lasting change will need to come from within each and every one of us. For more information visit www.amnestyusa.org/women/rapeinwartime.html.

Again, we are faced with real-life atrocities and if you follow the problem to a common denominator you will find a belief, and no matter how absurd, it's a belief that is believed in fervently and without compassion. Now I ask the question again: How is it that a person can come to believe something that is so obviously wrong and that can be evidenced by asking the believer if he would have his mother or wife subjected to this type of treatment? Again, the answer falls in the category of "neural networks" that have been set in motion and are self-limiting, letting nothing in except that which supports the preexisting belief. This is why people can be taken out of these situations, with the help of others come to see their insanity, and sometimes come to be the best opponents of such beliefs.

So my message to the women of this world is this: "It's time for change, for you, your daughters and your daughters' daughters. Don't 'believe' what you've read in your Holy Books because it was not written by the 'all-loving creator' of the universe. It was written by the same men who invented God as a man and in so doing, took all your rights away as a human being. It's also increasingly important for all of us, men included, to find balance with feminine energy within ourselves and whether or not men will admit this, we need women to be strong, so in the words of Bob Marley, "*Stand up—stand up for your rights.*"

CHAPTER 18

Some Things Never Change . . . Unless We Change Them

If I turn on the news it could be a broadcast from a year or two ago—it's the same thing over and over. As a matter of fact, it could be 40 years ago but instead of Vietnam it is Iraq and instead of Cambodia it is Iran and instead of Nixon it is Bush, but it's the same story. I know this is politically incorrect but to tell you the truth, I don't care about being politically correct anymore; we live in a country where everyone who is at least semi-informed knows there is something very wrong at high levels of government and that most of our national organizations are failing. The Katrina hurricane is a recent event that exposed the inadequacy and corruptness of our governmental agencies and politicians. The truth is that our government is politically incorrect, if not out-and-out criminal in a lot of cases. This is not just the isolated opinion of one person; it's the opinion of many senators, congressmen and women, and our representatives.

More recently, in June 2008, former Bush press secretary Scott McClellan told Congress under oath what he never said from the White House lectern—that top officials misled the country about Iraq and are still concealing the truth about leaking a CIA operative's name. Mr. McClellan said: *"The vice president has information that has not been shared publicly. You could go down the list, from Scooter Libby to Karl Rove, Ari Fleischer. There are others that have not shared everything they know about this."*

People are starting to speak out, and now we need people to listen and take action. Personally, I'm not the type of person to meekly take my poison for telling the truth and crawl off and die like Socrates, or to behave like Galileo, who abjectly renounced his own seminal discoveries under threat from the Church.

When you research the evidence you find there's little doubt that we have been led into every war, including the Revolutionary War and the Civil War, by large banking interests that do not have our "best interests" at heart. More recently the evidence surrounding 9/11 leads one to believe that it could have been an inside job to some degree, or at the very least the event was used to manipulate us into another war. If this is not true, it would be an exception to the rule but still leaves you wondering, "What is really going on and how do they get away with it?" That is what this book is about; they, meaning governments that are controlled by corporate interests or what's become known as the military-industrial complex, can easily manipulate a person to commit atrocities because they previously got the person to believe absurd beliefs . . . it's the same deficient neural networking for both scenarios.

Think of what has been happening during the past 50 years within our Christian religious institutions. It's only a matter of time before it finally comes out that the Catholic Church is all but bankrupt, both financially and morally. At the same time it has billions of dollars in a Rothschild's bank in England that no one seems to want to talk about. The reason I wrote "morally bankrupt" is apparent to most people but for those who are still in the dark, let me put it this way: I remember when I was 16 years old talking about the Church with my brother-in-law who was, and still is, a priest. The subject of molestation came up briefly and of course he denied it was going on. But as a 16-year-old Protestant I already knew that it was going on and I have wondered all these years why no insider within the Church ever came out in protest. This has been the biggest and worst cover-up in history of something that has affected thousands of families just in our generation. My point is simply this; if I knew, so did they and they said nothing, absolutely nothing—thousands of priests, nuns, bishops, archbishops, and the popes themselves fell silent and moved their offensive priests from one parish to another knowing full well what had happened and more important, what would happen again, and again, and again. And keep in mind that there are volumes of court testimonies, documentaries, and hundreds of millions of dollars in payoffs to keep parents and children silent. Just recently a Catholic church in Los Angeles had to pay $660 million to settle molestation lawsuits, bringing the total of such payments to more than $2 billion in the United States.

Of course we could also write volumes on the scandals within the Protestant religions. The Tammy Faye and Jim Bakkers, Jimmy Swaggart,

Ted Haggard, and the list could go on for pages.[52] What I find truly amazing about religious and political figures who are caught up in sex and drug scandals is that they are often exposed by the Richard Dawkins of this world, as seen in his documentary called The Root of All Evil, well before they are found out by their followers. This could very well be the best definition of "blind belief."

So it's no wonder that we have millions of Christians looking into Eastern religions for guidance and following gurus from other traditions. There's nothing wrong with this but the ironic part is that when you read Gnostic tradition and what Jesus says in some of the Gospels you find some of the most inspirational and loving thoughts ever written, such as loving your enemy, but I have yet to meet one Christian who follows that line of thought—not one. The idea of loving your enemy alienates most Christians right away. If it's true when Jesus says you can tell what kind of tree it is by the fruit it bears, you need not look too far or too long to see that something is not right about this tree.[53]

In the end, what's ironic is the fact that hidden deep in the inner teachings of some religions are transformative truths. In *The Power of Now* (1999), Eckhart Tolle writes:

> *When I occasionally quote the words of Jesus or the Buddha . . . or from other teachings, I do so not in order to compare, but to draw your attention to the fact that in essence there is and always has been only one spiritual teaching, although it comes in many forms. Some of these forms, such as the ancient religions, have become so overlaid with extraneous matter that their spiritual essence has become almost*

---

52 Richard Dawkins produced a documentary called *The Root of All Evil* (2007) about Ted Haggard and everyone who watched it knew that he was up to no good, strange, and would eventually be found out, yet, it took years for his flock to finally catch on. Why? Because they "believed" in him.

53 Matthew 7:15: Beware of false prophets, who come to you in sheep's clothing but inwardly are ravenous wolves. 16: You will know them by their fruits. Are grapes gathered from thorns, or figs from thistles? 17: So, every sound tree bears good fruit, but the bad tree bears evil fruit. 18: A sound tree cannot bear evil fruit, nor can a bad tree bear good fruit. 19: Every tree that does not bear good fruit is cut down and thrown into the fire. 20: Thus you will know them by their fruits.

completely obscured by it. To a large extent, therefore, their deeper meaning is no longer recognized and their transformative power lost. When I quote from the ancient religions or other teachings, it is to reveal their deeper meaning and thereby restore their transformative power—particularly for those readers who are followers of these religions or teachings. I say to them: there is no need to go elsewhere for the truth. Let me show you how to go more deeply into what you already have.

Again, I want to remind you that this is not an attack on religion itself, at least not directly. My issue is entirely with "beliefs" and "belief systems" that lock you into a mind-set with no way out because they are based entirely on faith and are no longer open for dialogue or further interpretation. It's just that religious beliefs are so easily recognizable because they are so dangerous and prevalent in today's world. That, along with the fact that they are so far out of date with proven scientific discoveries, makes them an easy subject to address.

I'm bringing this up because if you disagree or have closed down to anything I have said so far, you will have a very difficult time going any further because your mind and ego will filter out everything significant, important, and rational, and leave you in a conundrum. You could easily read this whole book and be untouched by any of the information and be convinced that I am nothing more than a fool. I'm including a list of resources in the bibliography that you can refer to, the reason being that I don't want to spend time trying to convince you that what I've said so far is true; you need to come to some basic conclusions for yourself before you are ready to read further. I have come to realize that a vast group of people will never read this book, not because they *won't* understand it or refuse to see, but because they *can't* understand it because of their limited neurological pathways of understanding and perception. For me to write a book on this subject and expect them to understand is rather like writing a song for deaf people to listen to; it simply isn't feasible or reasonable.

And even though you may occasionally feel the sharp edge of reason, this book is also written with love and compassion, with the knowing that we are all brothers and sisters on this tiny fragile planet and we need to work together through a better understanding of and respect for each other. Indeed, as Carl Sagan once said, *"We are probably going through something quite common on an evolutionary planet that has evolved to self awareness and are in our technological adolescence. It may be quite common for a planet to evolve to this level of technology and to quickly destroy themselves."*[54]

Our scientific power has outrun our spiritual power.
We have guided missiles and misguided men.
Martin Luther King Jr. (1929-1968)

---

[54] Carl Sagan was an American astronomer and astrobiologist and a highly successful teacher and documenter of astronomy, astrophysics, and other natural sciences. He pioneered exobiology and promoted the Search for Extra-Terrestrial Intelligence (SETI). He is world famous for writing popular science books and for cowriting and presenting the award-winning 1980 television series *Cosmos: A Personal Voyage*, which has been seen by more than 600 million people in more than 60 countries, making it the most widely watched PBS program in history. His book *Contact* (1997), which I highly recommend reading, was made into a movie shortly after his death.

CHAPTER 19

Religious Conflict That Is Happening Now

I'm sure there are many of you, including quite a few of my friends, who would say, "Well, things are getting better so let's keep on with business as usual and it will all work out." I believe this to be true, but not without opening our eyes to what is going on and why. Religious conflict will not go away by ignoring it. It's imperative that all of us look at what is going on in our world and trace the causes of these events back to their source, not for blame but for the inspiration to change things.

How to change paradigms: You keep pointing at the anomalies and failures of the old paradigm, you keep speaking louder and with assurance from the new one, you insert people with the new paradigm in places of public visibility and power. You don't waste time with reactionaries; rather, you work with active change agents and with the vast middle ground of people who are open-minded.
Donella Meadows, *"Leverage Points: Places to Intervene in a System"*
(The Sustainability Institute, 1999)

It doesn't take much research to track down the sources for legitimizing war crimes. Here I'll cite a few that can be pulled from the Bible and the Quran; you can easily see that adding divine sanctions to any dogma can be dangerous.

- Quran: 8:60: Prepare for them whatever force and calvary ye are able of gathering to strike terror into the hearts of the enemies, of Allah and your enemies.

- Quran: 47:4*:* Therefore when ye meet the unbelievers, smite at their necks and when ye have caused a bloodbath among them bind a bond firmly on them.
- Samuel 15:3: This is what the Lord almighty says: "Now go attack the Amalekites and totally destroy everything that belongs to them. Do not spare them; put to death men and women, children and infants, cattle and sheep, camels and donkeys."
- Numbers 31:17-18*:* Now kill all the boys. And kill every woman who has slept with a man, but save for yourselves every girl who has never slept with a man.
- Deuteronomy 20:16: However, in the cities of the nations the Lord your God is giving you as an inheritance, do not leave alive anything that breathes.
- Exodus 15:3: The Lord is a man of war: the Lord is his name.

Of course, it's easy to underplay this by saying that people don't actually take these passages seriously. But anyone who is in touch at all with our world would not say that. Let's take a look at a few things in our world that are a direct reflection of the divine sanctions of war.

- This was said by Lt. Gen. William G. Boykin, a high-ranking official in our military: *"We in the army of God, in the house of God, kingdom of God have been raised for such a time as this."*
- On the office door at Fort Riley*: "We should invade their countries, kill their leaders, and convert them to Christianity."*
- Sign above the door at the Official USAF Training Classroom at the Maxwell Air Force Base: *". . . they shall mount up with wings as eagles . . ."* (Isaiah 40:31)
- The purpose and vision on the Officers Christian Fellowship website: *"A spiritually transformed military, with ambassadors for Christ in uniform, empowered by the Holy Spirit . . ."*

Consider this: Since 1941 the United States has attacked, among others, North Korea, North Vietnam, Cuba, Cambodia, Laos, the Dominican Republic, Guatemala, Grenada, Panama, the Philippines, Libya, Iran, Somalia, Yugoslavia, Haiti, Sudan, Afghanistan, and Iraq. Not one of these countries was the aggressor or declared war on us.

If you want to really see religious insanity in action, look into the Campus Crusade for Christ and how embedded it is in our government. It stands behind something called The Great Commission.[55] If you think the Crusades are a thing of the past, consider this; we have more than 700 military bases in more than 130 countries and we are still building more with plans to invade more countries—sorry, I meant "liberate" more countries. It seems that it would be prudent to at least mention the small detail that it's against the Constitution of the United States *for religious groups, individuals, or organizations to engage in the machinery of the government.*

Another point that needs to be made clear is that if you follow this militant line of thought based on religious mandates, you end up invariably in an apocalyptic ending, and now there are numerous world governments that have the military strength to bring this about. Basically what the Campus Crusade for Christ and the Christian Embassy are about is the indoctrination of soldiers for the Christian agenda using the American government and your tax dollars to accomplish it. We do have watchdog organizations exposing groups such as the Christian Embassy and the Campus Crusade for Christ; you can go to www.militaryreligiousfreedom.com for more in-depth information and video clips that show clearly what is going on. A very revealing documentary called *Constantine's Sword* by James Carroll uncovers much of what is going on in places such as the U.S. Air Force Training Base in Colorado Springs. Organizations such as the National Association of Evangelicals, led by Ted Haggard, have put so much pressure on non-Christians at the Air Force base that there have been congressional investigations into the matter. Of course it didn't help the group's cause when Haggard was found to be having homosexual relations with at least one of the young boys in his congregation while high on methamphetamines.

If you really want to look at the social consequences of belief systems, consider the following statement of faith of the Christian Embassy. After you read this, consider the possibility of people with these beliefs pointing a gun at you or having their fingers on the button, which they do.

---

[55] Christ's command to convert whole nations with the aim of developing a worldwide Christian civilization and culture.

Statement of Faith
from www.christianembassy.com

The sole basis of our beliefs is the Bible, *God's infallible written Word* [emphasis added], the 66 books of the Old and New Testaments. We believe that it was uniquely, verbally and fully inspired by the Holy Spirit and that it was written *without error* [emphasis added] (inerrant) in the original manuscripts. It is the supreme and *final authority in all matters* [emphasis added] on which it speaks.

We accept those areas of doctrinal teaching on which, historically, there has been general agreement among all true Christians. Because of the specialized calling of our movement, we desire to allow for freedom of conviction on other doctrinal matters, provided that any interpretation is based upon the Bible alone, and that no such interpretation shall become an issue which hinders the ministry to which God has called us.

1. There is *one* [emphasis added] true God . . .
2. Jesus Christ is God, the living Word . . .
3. He lived a sinless life and voluntarily atoned for the sins of men by dying on the cross . . .
4. He rose from the dead . . .
5. He ascended bodily into heaven . . .
6. Man was originally created in the image of God.
7. Man's nature is corrupted, and he is thus totally unable to please God.
8. The salvation of man is wholly a work of God's free grace . . .
9. It is the privilege of all who are born again of the Spirit to be assured of their salvation from the very moment in which they trust Christ as their Savior.
10. The Holy Spirit has come into the world to reveal and glorify Christ and to apply the saving work of Christ to men.
11. Every believer is called to live so in the power of the indwelling Spirit that he will not fulfill the lust of the flesh but will bear fruit to the glory of God.
12. Jesus Christ is the Head of the Church, His Body, which is composed of all men, living and dead, who have been joined to Him through saving faith.

13. God admonishes His people to assemble together regularly for worship . . .
14. At physical death the believer enters immediately into eternal, conscious fellowship with the Lord . . .
15. At physical death the *unbeliever* [emphasis added] enters immediately into eternal, conscious separation from the Lord and awaits the resurrection of his body to everlasting judgment and condemnation.
16. Jesus Christ will come again to the earth—personally, visibly and bodily—to consummate history and the eternal plan of God.
17. The Lord Jesus Christ commanded all believers to proclaim the Gospel throughout the world and to *disciple men of every nation.* [emphasis added] The fulfillment of that *Great Commission* [emphasis added] requires that all worldly and personal ambitions be subordinated to a total commitment to "Him who loved us and gave Himself for us."

If you still need evidence of religious interaction with our military, check out Task Force Patriot-USA, whose mission statement is: "*Task Force Patriot exists for the purpose of sharing the fullness of life in Jesus Christ with all U.S. military, military veterans and families.*" I've talked with members of this group and I have to give them credit for the good work they are doing with veterans since our government doesn't seem to take care of them. Where I have a problem is when they directly engage with our active military in such a way that it gives the impression to the world that our military is a Christian military force, which brings up visions of the Crusades and Inquisitions.

So here are just a few conflicts that are either happening or have happened within the past few years on our planet. This has been going on for so long that we are almost immune to recognizing the underlying causes, which are predominately divisive beliefs based on race or a belief in God that makes your group special. The following information was derived from information sent to me by the Religious Tolerance Organization.[56] Believe it or not, there is actually quite a bit more than this, as far as conflicts going on, but this is enough to make my point.

---

[56] Religious Tolerance: www.religioustolerance.org

Country	Main religious groups involved	Type of conflict
Afghanistan	Extreme, radical fundamentalist Muslim terrorist groups and non-Muslims	Osama bin Laden heads a terrorist group called Al Quada (The Source) whose headquarters were in Afghanistan. They were protected by, and integrated with, the Taliban dictatorship in the country. The Northern Alliance of rebel Afghans, Britain, and the United States attacked the Taliban and Al Quada, establishing a new regime in part of the country. The fighting continues.
Bosnia	Serbian Orthodox Christians, Roman Catholics), and Muslims	Fragile peace is holding, only because of the presence of peacekeepers.
Côte d'Ivoire	Muslims, indigenous, and Christians	After the elections in late 2000, government security forces "began targeting civilians solely and explicitly on the basis of their religion, ethnic group, or national origin. The overwhelming majority of victims come from the largely Muslim north of the country, or are immigrants or the descendants of immigrants . . ." A military uprising continued the slaughter in 2002.
Cyprus	Christians and Muslims	The island is partitioned, creating enclaves for ethnic Greeks (Christians) and Turks (Muslims). A UN peacekeeping force is maintaining stability.
East Timor	Christians and Muslims	A Roman Catholic country. About 20 percent of the population died by murder, starvation, or disease after they were forcibly annexed by Indonesia (mainly Muslim). After voting for independence, many Christians were exterminated

		or exiled by the Indonesian army and army-funded militias in a carefully planned program of genocide and religious cleansing. The situation is now stable.
India	Animists, Hindus, Muslims, and Sikhs	Various conflicts that heat up periodically producing loss of life.
Indonesia, province of Ambon	Christians and Muslims	After centuries of relative peace, conflicts between Christians and Muslims started during July of 1999 in this province of Indonesia. The situation now appears to be stable.
Indonesia, province of Halmahera	Christians and Muslims	30 people killed. 2,000 Christians driven out; homes and churches destroyed.
Iraq	Kurds, Shiite Muslims, Sunni Muslims, and Western armed forces	By mid-2006, a small-scale civil war, primarily between Shiite and Sunni Muslims, started. The situation appears to be steadily degenerating. As of 2008 the degeneration has not let up and suicide bombers are a commonplace event.
Kashmir	Hindus and Muslims	This is a chronically unstable region of the world, claimed by both Pakistan and India. The availability of nuclear weapons and the eagerness to use them are destabilizing the region further.
Kosovo	Serbian Orthodox Christians and Muslims	Peace enforced by NATO peacekeepers. There is convincing evidence of past mass murder by the Yugoslavian government (mainly Serbian Orthodox Christians) against ethnic Albanians (mostly Muslim).
Kurdistan	Christians and Muslims	Assaults on Christians (Protestant, Chaldean Catholic, and Assyrian Orthodox). Bombing campaign under way.

Macedonia	Macedonian Orthodox Christians and Muslims	Muslims (often referred to as ethnic Albanians) engaged in a civil war with the rest of the country, which is primarily Macedonian Orthodox Christian. A peace treaty has been signed. Disarmament by NATO is complete.
Middle East	Jews, Muslims, and Christians	The peace process between Israel and Palestine suffered a complete breakdown. This has resulted in the deaths of thousands, in the ratio of three dead for each Jew. Major strife broke out in September 2000. Major battle in Lebanon during mid-2006. No resolution appears possible.
Nigeria	Christians, animists, and Muslims	Yourubas and Christians in the south of the country are battling Muslims in the north. Country is struggling toward democracy after decades of Muslim military dictatorships.
Northern Ireland	Protestants and Catholics	After 3,600 killings and assassinations over 30 years, some progress has been made in the form of a ceasefire and an independent status for the country.
Pakistan	Sunni and Shiite Muslims	Low-level mutual attacks.
Philippines	Christians and Muslims	A low-level conflict between the mainly Christian central government and Muslims in the south of the country has continued for centuries.
Russia, Chechnya	Russian Orthodox Christians and Muslims	The Russian army attacked the breakaway region. Many atrocities have been alleged on both sides. According to the Voice of the Martyrs: "In January 2002 Chechen rebels included all Christians on their list of official enemies, vowing to 'blow up every church and mission-related facility in Russia.'"

South Africa	Animists and "witches"	Hundreds of people, suspected and accused of being witches practicing black magic, are murdered each year.
Sri Lanka	Buddhists and Hindus	Tamils (a mainly Hindu 18 percent minority) are involved in a war for independence since 1983 with the rest of the country (70 percent Sinhalese Buddhist). Hundreds of thousands have been killed. The conflict took a sudden change for the better in September 2002, when the Tamils dropped their demand for complete independence. The South Asian tsunami in December 2004 induced some cooperation. The situation in mid-2006 is degenerating.
Sudan	Animists, Christians, and Muslims	Complex ethnic, racial, religious conflict in which the Muslim regime committed genocide against both animists and Christians in the south of the country. Slavery and near-slavery were practiced. A cease-fire was signed in May 2006 between some of the combatants. Warfare continues in the Darfur region, primarily between a Muslim militia and Muslim inhabitants.
Thailand	Buddhists and Muslims	Muslim rebels have been involved in a bloody insurgency in southern Thailand—a country that is 95 percent Buddhist. The army has seized power and has agreed to talks with the rebels.
Tibet	Buddhists and Communists	The country was annexed by Chinese Communists in late 1950s. Brutal suppression of Buddhism continues.
Uganda	Animists, Christians, and Muslims	Christian rebels of the Lord's Resistance Army are conducting a civil war in the north of Uganda. Their goal is a Christian theocracy whose laws are

based on the Ten Commandments. They abduct, enslave and/or rape about 2,000 children a year.

With this "incomplete list" of religious conflicts going on we must admit that there are *severe social consequences* of belief systems that divide us. I think at this point you would agree that being a religious moderate that tacitly supports extremist organizations does count when figured into the equation. It's easy to gloss over some very fundamental and important facts; for instance, more than 150 million humans have been killed by other humans in the 20th century alone, and this was accomplished to a large degree by people whom we would call normal and good upstanding citizens. And keep in mind that this was accomplished with weaponry that pales in its inefficiency when compared to 21st-century weapons developed by technology.

The dysfunction of the human mind is what has created the problems we see in the world today.
Eckhart Tolle

We must deeply consider the above quote as we allocate billions of dollars for the military-industrial complex to build "peacekeeping weapons."

CHAPTER 20

Immaculée Ilibagiza

On Forgiveness
An Interview with Immaculée Ilibagiza

I have often wondered, "What would I do if someone killed someone I love?" In today's world a lot of the violence is in the name of revenge. You need to go back decades or centuries, or in the case of the Middle East, millennia, to find the original insult or wrongdoing. In the case of the Middle East it's, as Bill Moyers of PBS said, "The longest-running family quarrel in history." It's been so long that most Israelites, Christians, and Muslims have forgotten what it was even about, yet the conflict goes on, and on, and on. That's why, when I watched Immaculée on *Oprah* and on *60 Minutes* and I read her books *Left to Tell* (2006) and *Led by Faith* (2008), I knew I needed to talk with her to see how she came to that place of forgiveness of the people who slaughtered her family, friends, and neighbors in Rwanda. The time has come when we must learn to forgive the "others." Unless we do, there will always be a reason and justification for revenge and the fighting will continue. Fortunately it can no longer continue on our planet the way it has for the last 3,000 or 4,000 years because living in a world with high technology that is ruled by "tribal warfare" is no longer an option. And when I use the term "tribal" I am not limiting it to the traditional meaning of the word. I have witnessed tribal warfare by those who drive tanks and Cadillac Escalades.

I thought it was important to show that there are aspects of faith and how it can be used that are useful if not necessary in today's world. It's only when faith is used as a reason to no longer question our beliefs that I see a problem, not as it is used to get us through the tough times in our lives. When all is lost and seemingly hopeless, faith may be the only thing left to us.

I hope you get out of this interview as much as I did.

Rahasya: Some of my questions may be on a deeper level than what you are used to because I happen to be working on forgiveness myself.

Immaculée Ilibagiza: I think we always are.

Rahasya: After reading your book,* Led by Faith, *and trying to understand all you have been through I barely feel qualified to ask some of the questions I have because I have never experienced anything even close to what you have.

Immaculée Ilibagiza: Please ask anything you want.

Rahasya: I think to help explain where you are today we need to have a brief background of your experience in Rwanda. I think it's difficult for a lot of people to understand the depth of the hatred between people who were for all appearances the same. We are used to that when dealing with obvious differences like skin color and religion, but

this seemed to come from an intentional design on the part of other groups of people with their own agendas from other countries.

Immaculée Ilibagiza: Yes. What happened in Rwanda started with colonialization. The first group were Germans who colonized the country; then they gave the country to Belgium. When the Belgians came they found a country that was very organized. It was a surprise for them to see a country in Africa that was so organized, with a king and every area had a chief. And the king was a Tutsi. Anyone who had more cows, which was a sign of wealth, would become Tutsi. If a Tutsi loses a cow, he would be poorer; he would become a Hutu. Like any other country, there are the rich, and the middle class, and the class that does not have much money is always the largest. There wasn't much of a noticeable difference between the Hutus and the Tutsis. Sometimes even between two brothers, one of them would become a Hutu and the other become a Tutsi. We're now looking at the consequences it can have on a country when it divides itself, no matter what the division is based on.

So then in a very short time the Hutus overthrew the king. Many people believe that he was poisoned. And then they trained the Hutus to take over. The people who took the power in the first government were people who didn't even know how to read sometimes. They just wanted to have the power so they could overthrow the Tutsis. What was once a class division had become a political division. So there was a big conflict between the tribes. So the Hutus had one agenda, and it was to kill Tutsis or send them away. This was the first civil war in 1959.

The second war was in 1967. In 1973 they made another war against Tutsis who were going to school. In 1994, it was again really the leaders who arranged it. And to this day, I don't think there can ever be genocide if the leaders are not behind it. People don't hate each other enough to get up and start to kill each other. You must have strong people behind it.

Rahasya: Yes, they've done a lot of research over the years; it's the "madness of crowds." There is a crowd psychology, like a lynch mob, for instance. Most people in a lynch mob would never lynch a person if they were by themselves, but in a mob, or a crowd, you can hide.

And we do some of the same things in today's world. Some of our belief systems, our nations, and more recently our corporations can turn into a mob mentality.

Immaculée Ilibagiza: That is so right. The important thing I would like to say is that if it could happen in Rwanda it can happen anywhere there are strong leaders behind it. People that want power so badly end up having evil in their hearts. When you see firsthand something like this happen and understand the process behind it, you should realize it can happen in any country. You can't believe that it happened to this beautiful country. People say to each other, "How can this happen right here?" And those who took part in the war are now regretting it.

Rahasya: There was a teacher who did an experiment a few years back, and she divided her class into blue eyes and brown eyes, just for an experiment, and by the third week, she had to stop the experiment because they started fighting on the playground; the blue eyes started making up stories about the brown eyes—they weren't as smart—not as strong. There have been a series of experiments and studies that point to the fact that we have a mental process that strives to divide. Some seem to think it is part of our evolution that helped us survive. And the only way we can get past this is doing what we're doing right now, which is dialogue, talking and putting light on it.

Immaculée Ilibagiza: Yes, once you acknowledge the conflict and put a little light on those things that go on in the mind, you can feel it coming and make a conscious choice to take this action or that. One problem was the fact that we couldn't study the history of such things as the Jewish Holocaust; it was forbidden.

Rahasya: I think that with everything going on in the world today, and with the position that the United States is in with so many conflicts in the world, we're taking a deep honest look at ourselves here in the United States, finally, and to do that, I think we need to figure out this next question. According to some research, three out of five conflicts are motivated by revenge. The only thing that seems to heal this is forgiveness. Can you go into what you went through in the process of forgiveness? How did you "truly" forgive, Immaculée?

Immaculée Ilibagiza: Oh, that's such a good question. I remember in the first week of hiding, forgiveness was not even in my reality; I was so angry and scared. In my life up to that point I had my parents, my brother, I had people loving me. I never knew anyone hating me in my life. I thought everyone was so honest in life, even the government. In that kind of world I didn't really see or understand all that was going on around me. And all of a sudden I am sitting in the bathroom and I'm thinking, "What will happen to me? What has caused this to be this way?" So that week I was more just realizing the reality in the world: "What is going on?" I was just lost. I was so overwhelmed with this anger and thinking, "If I get a chance to live, I will never speak to Hutus, I will hate them and their children, and I would never be friends with them after what they did to me, my brother, my mother, my dad." All I could think of was "these evil people." I hated them, thinking they were not humans. This was all real. I was so angry. And then the pain has a way of teaching us, even though we hate it. And I remember sweating out of anger. Just trying to imagine what I will do when I get out. I will put the bomb under the country and everyone will blow up. I can just be like Rambo and I can shoot everybody. I wanted to be anything horrible, terrible, and kill these people.

I found that there are *steps* to forgiveness. The first step was to *feel the anger.* To really feel it as it is, in your body. And then I remember when it was really painful, my body was aching, I couldn't even remember how to smile. I wish I could be naive and innocent again. Smile to people, be happy to people. Believe in the goodness in them. But right now I only believe in their evil. I knew I had to let it go. I was thinking to myself that the anger I had toward the Hutus was so much, was so heavy, I didn't know what to do with it. And it was aching my body. And to tell the truth, there was one time I thought I would never know how to forgive and a part of me that didn't want to forgive them. I have such a good reason to hate them. Finally I had to realize that we are all children of the same God.

Rahasya: Yes, there's a part in the Bible where it says, "Why would my right hand cut off my left hand?" We are all one humanity. It looks like we're divided but energetically, we're like this thin film of consciousness, a planetary consciousness, in a planetary body, and I think the Bible

speaks of it as being one in the body of Christ, Pierre de Chardin speaks of the noosphere surrounding the earth, and progressive spiritual thinkers have been speaking of "spiritual evolution" for centuries. But the reality is that you and I, speaking right here, right now, we are the same person, speaking to ourself, from this higher consciousness that is evolving and insisting on unity and oneness.

Immaculée Ilibagiza: That is so true. And that is really the lesson you get there, let go of that voice that tells us that we are different. Not only are we praying for peace in the world, we also need to pray for peace within our own lives. And you don't have to take care of everybody. But send them love. Think of them as people who need to express themselves, who are trying to become. When you forgive, there are so many people who feel it, that know it, and you have let go of the anger toward other people. Instead of them now continuing to build up the anger of the revenge, or what they can do to you, and your loved ones, you give them a chance to think and to feel their hearts themselves. If you hold anger they have that passion of continuing to be angry with you, then they are trapped and live with their fear of what you can do to them. But once you can forgive them in the depth of your heart, something happens. And the great thing about it is they will tell someone, maybe their child, their wives, their friends. Maybe not what they have done, because it's not good. Then you give them a chance to grow. And you don't have to go to save them, because these feelings of love, we all know it and love does its own work.

Rahasya: Yes, this has to be the most important lesson, and maybe our ultimate lesson on this planet, as we grow into a global society, because in today's technological world you can only imagine what would have happened in Rwanda if the Hutus had access to fighter jets with advanced technology or an atom bomb. So we really need to get this under control, this armed conflict; our barbaric past has to come to an end. Otherwise, there's going to be one person who will bring it all to an end because unlike even 200 years ago, it only takes one person to kill a million people now.

Okay, I know this is a politically incorrect question but being politically correct is one of the things that have brought our planet

to the brink of destruction so . . . I'd be really interested to hear what you have to say about this. According to Amnesty International, there's usually an average of 20 religious conflicts going on in the world at any one time. These are armed conflicts by people of "faith" who have a belief in a loving, although capricious and jealous, God. More often than not they are of the same skin color, the same genetics, and living in the same area; the only thing that divides them is their beliefs. What have you learned from your experience in Rwanda that you feel would benefit them? In other words, people of belief who have the belief that they have the ultimate truth that gives them the ordained right by their God to change other people or remove them from the planet.

Immaculée Ilibagiza: Oh, my God. I wish I could have a quick answer to that, you know, one thing that would completely answer that question. But you know what I think, if you really understand what love is and what it does to you, you really see yourself in other human beings, and try to act with love. I wish there could be a government of love. A ministry of love. Somewhere where people have the business of being compassionate. The business of knowing what is going on in the heart of another person. To just hear what you think and what you are feeling. I just want to tell you the journeys I went through, the struggle I went through in my heart by hiding, the tears I shed for my mother, for my father, what would happen if you went through that. And I really think that if there is one thing that gives me peace in life, and which allows me to act as who I am, I can come to this country and feel like I have peace in settling here, is that knowing that deep down, we are the same. Deep down we have the skin that gets hurt the same way, deep down we like people who talk to us with love, and we feel bad with those who talk to us harshly, and we can pretend it doesn't touch us, but yet it does. That is how you can defend another person who is being treated unjustly. Because even if you don't know them, you know those things hurt. When we believe in something, you can't critique somebody else for believing what they believe in, it's a belief, which means you don't know for sure. I used to have a boyfriend, he was from another region, all Christians, but they had different beliefs. And I had a different belief in one small thing but I accepted him as he was but he wanted me to change if we were going to get married. And I loved him with all my heart, and I wished I could change. But what does it serve you if I change something, just for you, and not because I'm really convinced? And is that really change?

Rahasya: Yes, because usually what we're asking another person to do is to pretend like they believe something that they really don't believe. I think people forget that to believe something it needs to be believable. Bertrand Russell, a great Christian writer, once wrote that he would never die for his beliefs, because he could be wrong. And if we can all grasp that, that all of our beliefs, almost traditionally and historically, almost every belief that we've ever had on this planet has ended up either being wrong or misleading and if we look at our beliefs now, so much of it is changing and in constant evolution; to get attached and identified to some fixed idea of what the reality of the world is, is really absurd and possibly a delusion of grand proportions. We need to hold on to that part of ourselves that we're children and constantly learning.

Immaculée Ilibagiza: That is why I love children. They are so innocent. They touch the fire, they let it burn them, and they learn from that, and what is most important is how we practice love. I have seen people in my religion, in Rwanda, killing those who are teaching us to love. So it's really how you grasp with your heart; it's not how much you know. It's how much you can practice that love. I will write again and again, you can never go wrong by caring for somebody, you can never go wrong by giving, by caring, by loving, but you go wrong by hurting somebody, by lying to somebody. You can go wrong by hating somebody. But if in your heart you are doing what you think you wish to be done to you, to another person, you can never go wrong. And I think that is the secret. How can we get there?

Rahasya: Well, like most truths, they're usually simple. All beings are searching for one thing, to love and to be loved and to connect with other beings doing the same. And I think that's about as simply as you can put it.

I thought it was important to include this even though there's a lot in the interview that I would disagree with. But my goal is not to simply go out and find people who agree with the way I see things. The goal is to find the best part of ourselves that exists in everyone and even though Immaculée uses religious terminology that I may not use, it was her process of forgiveness that I found compelling. I have found that the best answers come from the extreme experiences that we go through in life and it's rare to find people such as Immaculée who have had those experiences and lived to tell how they survived the hate those experiences gave rise to.

CHAPTER 21

Dan Millman

Being a Peaceful Warrior An Interview with Dan Millman

I'm sure many of you have read *The Peaceful Warrior* by Dan Millman or watched the recently released movie of the same title.[57] I thought it would be interesting to get Dan's ideas on being a peaceful warrior in today's world and what his mentor, a man he calls Socrates, had to say about it. If you haven't read or watched the movie be sure to catch it because it is filled

[57] You can find out more about Dan Millman by visiting www.peacefulwarrior.com.

with little gems that stay with you. He recently sent me his latest book, *Wisdom of the Peaceful Warrior* (2006), which I also highly recommend.

Rahasya: In* The Peaceful Warrior, *you say that Socrates's ignorance is based on the understanding that life is a mystery. And you, Dan, your understanding is based on ignorance. Could you elaborate on that and how this might apply to our world today?

Dan Millman: Well, Socrates had a way of making some outrageous statements to shake me up. If we were talking about how the concept of a peaceful warrior relates to the state of the world today, I would point to my most recent book, *The Journeys of Socrates*, the life story of my old mentor—how he became a warrior, but more important how he found peace. There are many people who have a peaceful heart, but who haven't yet found the warrior spirit. This is the paradox of the "peaceful warrior," the integration of opposites. It all occurs in the context of Mystery—the understanding that even with all our sophisticated knowledge, we are like children gazing at a night sky, blissfully ignorant of why we are here or where the universe begins or ends. As pundit H. L. Mencken said, "We are here and the time is now; all other knowledge is moonshine."

I believe that one of the major hurdles in the world today is this stubborn idea of an "us" and a "them," a competitive, rather than collaborative, approach to living. The Palestinians, the Israelis, the Kurds, Shiites, and all the various factions in Iraq. We could go on and on—skirmishes between India and Pakistan, Russia and China. At some point this world is going to have to begin to see that we're all in this together—a human family.

Rahasya: It seems we are on the cusp of a mysterious transformation on this planet and what you're talking about could very well be the predictor of whether we survive our technological adolescence and to evolve into a more cooperative spirit.

Dan Millman: Yes, some people say humans are always going to be fighting, but I don't believe that must be. As Socrates says in the *Peaceful Warrior* movie, "I call myself a peaceful warrior because the important battles we must fight are on the inside." When we find peace within, we will create it in the world. Violence reflects our current stage of evolution. When more human beings have enough to eat—when they have a sense of stability, security, meaning, and self-respect—they will have more to live for and less to die for. When we begin to see ourselves as one human spirit, then fighting and killing others will seem as crazy as our arm trying to hurt our leg. All one body, you see.

Rahasya: It's that kind of survival thinking that keeps us making decisions based on fear, which always leads to conflict.

Dan Millman: We have to walk before we run. As psychologist Abraham Maslow wrote, we have to satisfy one need before we can face another. Gandhi once said, "To a starving man, God is bread." Our first priority must be to bring support to those parts of the body of humanity that need attention, to encourage self-reliance and dignity. It is not the whole answer, but it is a good place to begin.

The answers don't lie exclusively in the east or in the west; it isn't a matter of pursuing the flesh or only turning to spirit. The way that I teach—and what is expressed in both my book and the *Peaceful Warrior* movie—is about balance—taking the best of east and west, flesh and spirit, science and mysticism, men and women. No one is smarter than all of us.

Rahasya: In one part of your book, you talk about living life spontaneously by being in the moment instead of reacting, and those reactions being determined by your past. Today we live in a world that's heavily invested in maintaining the old paradigm and old beliefs. What do you think it's going to take for all of us to let go and be transformed, or as Socrates would say, "disillusioned"?

Dan Millman: The theme of living in the moment is quite important, along with the reminder that there are no ordinary moments. Being in the moment really is all we can do, but we persist in believing that the past and future are real, without realizing that the past is a set of impulses in our brain we call memories and the future is simply our imagination. All we have, the only moment of reality, of sanity, is this moment.

Socrates first pointed out to me that "disillusion"—what we normally consider a negative experience—actually means a "freeing from illusion." Not always a pleasant experience, since we love our illusions, but an important leap forward, each time we see ourselves as we are.

The biggest illusion most of us cling to is the illusion of who we are. Many people know the value of self-acceptance, but we can only accept ourself if we know ourself. So, along the peaceful warrior's path, at some point we must face our own shadow, those aspects of our psyche that we disown and deny. Seeing ourselves realistically, with our strengths and our flaws and foibles, makes us less defensive, more compassionate and empathic with others. And we begin to see the world and other people more clearly.

Rahasya: Do you ever get the feeling, Dan, that if you were in anyone else's body on this planet, and closed your eyes, and said, "I am," you would have the sense it would be the same "I am" as you are right now?

Dan Millman: Yes! But that is part of the great paradox that Socrates always pointed to. On a conventional level, we are separate, individuated beings, responsible for our lives through free will and choice. But from a transcendental view, a higher truth informs us that we are that Witness, that same Awareness shining through all the billions of eyes of humanity. So I don't actually experience a separation from anyone. This understanding can restore our perspective and humor when we realize that all this drama, all this play of light and shadow, is the many and the One. And in this moment I look around and see the beauty of the world, but there is no inside and no outside, no me and no other. Just the One.

Rahasya: It seems when I get in trouble it's the same place where the world gets in trouble. We all need egos to communicate and deal with the world but we need to also keep in mind that it's not the real "I, that I am." My ego comes into existence through feeling separate and it does that by setting up boundaries and nothing makes those boundaries seem more real than conflict, hence on a national level we have wars.

Dan Millman: The ego gets a bad rap in spiritual circles. If someone says, "You have a big ego," that's not normally considered a compliment. But I use my ego, my personality level, this sense of identity, to teach, to learn, and to serve. When someone says, "I'm going to get rid of my ego," I have to ask them, "Who just said that?"

Part II:
The Neurological Consequences

CHAPTER 22

Neurological Consequences and Seeing through the Illusion of Life

Something I have noticed through the years about devout religious believers, or people who have strong ideologies such as Stalinism, Maoism, fascism, and extreme right Republicanism, or what some may call neoconservatism, is that they all seem to share a shortsightedness that borders on mental illness at times as far as how they see themselves and their relationship to and responsibilities in the world around them. They seem to live in their own reality and inevitably, at some point, it comes crashing down around them, and us. While talking with them at times I found that there clearly were some terms, concepts, and views they simply would not, or more specifically, "*could not,*" understand. This is what I'm calling the *neurological consequences* of belief systems.

If you would be a real seeker after truth, it is necessary that at least once in your life you doubt, as far as possible, all things.
René Descartes (1596-1650)

This is what we need to look at; why is it that people with strong beliefs look and act as if they are out of touch with the very reality that surrounds them, a reality that is in direct conflict with those same beliefs? And why is it that some of us believe and others don't? And why is it that believers are so against questioning their beliefs; it seems to us nonbelievers that they would want to question their beliefs and come up with some evidence for their validity. Of course we know why they don't want to question their beliefs, and we will deal with this later when we hear what people who have given this a lot of thought and research have to say.

My hope is that this book might actually help transform a person's consciousness in the same way my consciousness has been transformed through the years; it's been transformed in small degrees through maintaining an open mind and evaluating all the evidence with intellectual honesty. I hope this is a book that would actually change the way a person sees the world. I find it strange that I would be looked at as a skeptic or a heretic at this point in my life because it seems as though everyone should question his or her beliefs, but the hallmark of being a "believer" is the fact that believers never question their beliefs or how they came about in the first place. This book is my questioning of some of the beliefs we have about ourselves and the world we live in. But to do this takes two; you will need to enter into this with courage and trust because you may lose something that you hold very close and dear—your beliefs. I'm also talking about a shift of consciousness so slight that at first it's imperceptible but in the end has deep ramifications on how you view the world and more important, how you view your relationship to the world and to other human beings. A monk who, after many years of discipleship to his master, once asked, *"Master, after all this time with you I still sense a profound difference in us; what is it?"* The master answered, *"There is only one difference, you see yourself in the world, whereas I see the world in myself."*

There is nothing like returning to a place that remains unchanged
to find the ways in which you yourself have altered.
Nelson Mandela

In my life I have come across several important pieces of information that have opened my eyes in such a way that they have changed my worldview from the micro world to the macro world, and most definitely my inner world. Once your eyes are opened, you see the world as being interconnected and yourself as one with it. The sacred emerging reality that is happening is a worldwide phenomenon and it is affecting all of our outdated and dysfunctional organizations worldwide.

The information in this book needs to unfold gradually
so be patient with me and with yourself.

The Limitations of Our Perception

To let any new knowledge in it's also important to maintain a degree of humility. Arrogance and pride can blind us to the most obvious facts. When I first came into contact with the following information it was humbling to say the least, especially when I found myself extrapolating too far beyond the limits of my understanding.

Something to consider is the limits of what we actually perceive. Imagine for a moment that the electromagnetic scale was represented by the volume of the Empire State Building in New York. Now imagine that it was full of sand. What we perceive is the equivalent of one grain of sand and the rest is the reality that we cannot perceive. Now take that one grain of sand in your hand and keep in mind that we are perceiving only three dimensions of at least 11 dimensions that we know probably exist. And now we know that the matter that we perceive is only 5 percent of the matter in the universe; the rest is dark matter that we don't see at all.

So what we perceive is so small that to even understand how little we perceive is difficult to perceive. This isn't even taking into account the idea of multiple universes, parallel universes, and string theory.

If we accept that this is the miniscule amount of reality that we perceive, it becomes much easier to understand why and how there can be other beings and forms of consciousness that we are still not seeing, although our technology is now picking up forms of life energy that we can not normally see. For instance, Kirlian photography has been around for many years and more recently our digital cameras have been taking photos of orbs; I have seen this for myself and the best photo experts in the world can not explain it away as due to technical anomalies.[58]

So when formulating ideas, be careful if you find yourself coming to any kind of conclusion because we are not getting the whole picture, so to speak.

If the doors of perception were cleansed,
everything would appear as it is—infinite.
William Blake (1757-1827)

[58] Read *The Orb Project* (2007) by Miceal Ledwith and Klaus Heinemann, PhD.

Real information is catalytic in its truest sense and is meant to bring about a change in all that it touches. My greatest wish is for this information to be that catalyst for you to see the world as it is, not as you believe you are. We are approaching times on this planet when we must be clear and free from beliefs and superstitions that alter and limit our worldview and our conception of who we are in relationship to it. Most of us are blinded by layers of illusions, one on top of another. We define ourselves with layers of personalities within the illusions, but deep within, there is a sacred reality that is our home and we are that sacred reality. But we must learn to see the illusions so we can see through them, so in a very real sense I hope to disillusion you. Even seeing what is plainly before our eyes can be difficult or even impossible sometimes.

Consider the following visual illusion (from Roger Shepard's "Turning the Tables"):

The tabletops above are exactly the same in both size and shape. Go ahead, take a ruler, prove it for yourself, and then notice that they still look different. Our minds are hardwired to make certain interpretations based on special cues. If this is so with the visual cortex in our minds, we must at least entertain the possibility that we make the same type of mistakes when we mentally try to make sense of our existence. This is why it is imperative to maintain a very high standard of research, validation, and questioning while looking into these matters.

Another quick example of how easy it is to fool the senses is the next picture. There are no curved lines here, yet there appear to be curved lines. The way we are wired and what we expect to see determine to a large extent "what" we *do* see.

No Straight Lines

Here's another example that is more to the point. How many times have you seen this logo on the side of a truck? Most people have seen this hundreds, if not thousands, of times yet they have never seen the arrow between the "E" and the "x." The point isn't that they haven't seen it; the point is that once you have seen it, you can't help but see it all the time. In fact, it jumps out at you.

What I'm referring to is much like that; once your eyes are opened you can not help but see through the veils of illusion. You see through the reasons of so-called righteous men, you hear what is really being said by the "honest politician," the news is no longer one event after another—you see the process behind the events and easily connect the dots. With spiritual awareness and courage, not fear and trembling, you start to ask the difficult questions that so many people have traditionally run from. With eyes open you see clearly what wise men and women who have been persecuted throughout history have been saying all along, which is that we are and have been continually manipulated by organized religions, corporations, and governments.

So long as the people do not care to exercise their freedom, those who wish to tyrannize will do so; for tyrants are active and ardent, and will devote themselves in the name of any number of gods, religious and otherwise, to put shackles upon sleeping men."

Voltaire (1694-1778)

So continue if you dare.
All you will need is a light heart,
an open mind, and an adventurous soul.

CHAPTER 23

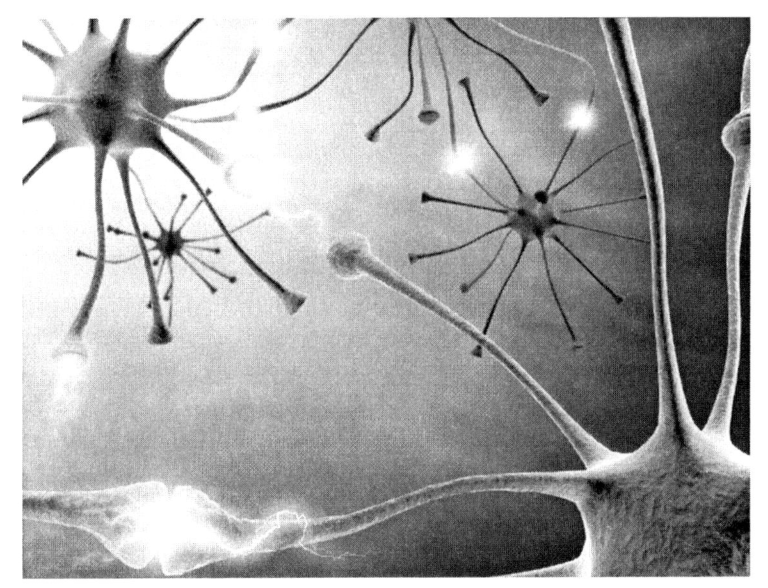

Neural Networks

Setting Up Neural Networks

When I started noticing that people with deep-seated beliefs had difficulty understanding certain concepts and ideas, I started wondering why or, more accurately, *how* could this be? This was where my search started for evidence showing that having any rigid belief such as a *religious dogma*, *strong ideology*, or *limiting personal beliefs* could possibly set up neural networks in our brains that could inhibit our conscious evolution and indeed, even our everyday understanding of life and nature by filtering and compartmentalizing insights and knowledge and even blinding us in the face of hard evidence in order to sustain a particular belief.

I need to remind you that I'm not referring to the power of believing in something such as having better health or living a life filled with abundance. Throughout history we have observed this in action and it's been written about by people such as Louise Hay, Napoleon Hill, Andrew Carnegie, Wallace D. Wattles, Stephen Covey, and hundreds of others. You will be reading my interview with Lynn McTaggart on the *power of intention.* The difference here is the fact that we are validating results and are open for dialogue and controversy without getting defensive and even violent. It has been shown that simply being open and seeking other points of view increases our neural connections and maintaining a belief inhibits those same connections.

A well-known term in neurology states, *"Neurons that fire together, wire together."* This means that if you increase your ability to learn new things, you also increase the number of neural connections, which in turn implies that if you have and maintain a limiting belief (and most beliefs are limiting) you set up a neural network that eventually inhibits new learning by breaking down neural connections because they no longer fire together. This is partially due to our ego personalities' identifying with our beliefs and seeking consistency by not changing.[59] This is also one of the big problems with addiction; you get stuck in a pattern that's all but impossible to get out of and addictions come in many forms and affect us on many levels. Belief itself is an addictive mental pattern in which you think the same thought over and over by pretending you know it to be true until you believe it to be the truth. This is also the definition of a trance. This isn't even considering the implications of being addicted to the chemistry attached to the emotions that fuel most beliefs. Every emotion we have has a chemical counterpart to which we acquire an addiction because the molecules of emotion fit like a lock and key into the opiate receptors in our cells. We've all known those individuals who seem to be addicted to feeling sorry for themselves. When this happens their subconscious minds will set them up in situations that will be conducive for setting those emotions in play; hence, we hear them say things such as, "I knew this was going to happen to me." Of course they knew it was going to happen; they planned the whole event, albeit

---

[59] This is what makes it possible to raise $27 million to build a museum that teaches our young children the world is only 6,000 years old to simply reinforce and continue previous beliefs. The museum is in Petersburg, Kentucky, and its website (www.creationmuseum.org) says, "Prepare To Believe."

unconsciously.[60] The same is also true when we protect our religious beliefs by filtering out all possible conflicting information so as not to upset our chemical balance.

We all have this tendency to set up mental patterns or strategies of transforming something that started out being something we believed *may* be possible to something that becomes an undeniable truth. Once we "believe" we know the truth we no longer search, question, or inquire, which means, as I stated before, that we start breaking down neural connections and setting up rigid and limited neural patterns that support only the belief. What this accomplishes for the "believer" is to create a mind that eventually is incapable of questioning its own beliefs because it no longer has the mental capacity (neural networks) to do so. This is a very dangerous situation, especially when it reaches the worldwide proportions that it has today with so many believers in control of our destiny. What was once only probable is becoming inevitable if we don't change our worldview by questioning our beliefs.

Men never do evil so completely and cheerfully as when they do it from a religious conviction.
Blaise Pascal (1623-1662)

Even though I am bringing up a lot of seemingly different issues, be assured that they are all linked to one thing, which is the way we believe our beliefs unquestioningly and how that affects our ability to evolve consciously and create a sustainable environment and society. I also see this particular level of brain functioning as necessary if we are to ever build a better world.

---

60 An excellent resource for following up on this idea would be Candace Pert, neuroscientist, and her book *Molecules of Emotion: Why You Feel the Way You Feel* (1997). Scientists have found neuropeptide receptors throughout the nervous system, and Pert's research has shown that the immune system also produces its own. She has come to believe that the brain and the nervous, endocrine, and immune systems are interlocked in a "psychoimmunoendocrine" network that serves as a multidirectional, bodywide system in which every part communicates with every other part. This concept nullifies the prevailing idea that the mind has power over the body. "Instead, emotions are the nexus between mind and matter, going back and forth between the two and influencing both," says Pert. This adds evidence to statements such as, "As a man thinks in his heart (emotions), so is he."

Breaking the Spell

We should also look at the other side of this question of why we believe. Is "believing" a part of our nature that's necessary to give us hope? By questioning the process of believing, will we break the spell that is necessary for mystical or religious experiences? I think not, but that's my belief and it also needs to be questioned and thought out. But it's disquieting to think that the continuation of our society depends on living in a mass delusion and that our survival will depend on singling out those who start to awake from the madness and persecuting or crucifying them into quiet submission.

Philosophy is questions that may never be answered.
Religion is answers that may never be questioned.
Anonymous

In the words of Daniel C. Dennett from his book, *Breaking the Spell:*

> *It is high time that we subject religion as a global phenomenon to the most intensive multidisciplinary research we can muster, calling on the best minds on the planet. Why? Because religion is too important for us to remain ignorant about. It affects not just our social, political, and economic conflicts, but the very meanings we find in our lives. For many people, probably a majority of the people on earth, nothing matters more than religion. For this very reason, it is imperative that we learn as much as we can about it.*[61]

As Dennett brings out in his book, billions of people worldwide pray for peace but the underlying intention is to spread peace through their own religion, which obviously has never set well for people of other religions. Meanwhile, the rest of us want to "know" the truth, not simply have a vague and oftentimes childish belief. It seems that "believers" would be standing in line to put their beliefs under the microscope of reason and knowledge to finally know for sure. I for one would be the first to join their cause if it

---

[61] *Breaking the Spell: Religion as a Natural Phenomenon,* 2006, by Daniel C. Dennett, professor of philosophy and codirector of the Center for Cognitive Studies at Tufts University.

were shown to be true, or even beyond a reasonable doubt. Of course when religious beliefs are put to the test of reason they are almost always shown to be false or at best, misleading.

The whole problem with the world is that fools and fanatics are always so certain of themselves, but wiser people so full of doubts.
Bertrand Russell (1872-1970)

CHAPTER 24

Dr. Andrew Newberg

How God Changes Your Brain
An Interview with Dr. Andrew Newberg

The first time I heard what Andrew Newberg had to say was in the movie/documentary *What the Bleep Do We Know?* (2004). The next time was in Bill Maher's *Religulous* (2008), which is a documentary on the ridiculous aspect of our religious beliefs. I was actually surprised by Bill Maher's documentary because it portrayed religious fundamentalism in a way that may reach a lot of religious believers. I wanted to hear what Andrew Newberg had to say on the topic of religious beliefs and the brain because of his background and the fact that he is very scientific when it comes to his ideas and conclusions.

Dr. Newberg has actively researched and documented his findings using neuroimaging in the study of neurological and psychiatric disorders. Some of his latest research focuses on how brain function is associated with various mental states, in particular, the relationship between brain function and mystical or religious experiences. This research led to his book *Why God Won't Go Away: Brain Science and the Biology of Belief* (2001). He teaches two undergraduate courses at the University of Pennsylvania—Science and the Sacred: Neurotheology, in the department of Religious Studies, and Imaging the Human Mind in the Biological Basis of Behavior program. His list of academic achievements are extensive and you can find out more about him on his website at www.andrewnewberg.com.

So you can understand why I wanted to interview him when I heard him make the statement, *"Fundamentalism, in and of itself, is benign and can be personally beneficial, but the anger and prejudice generated by extreme beliefs can permanently damage your brain."* As it turned out we didn't totally agree on everything. He is of the opinion that moderate religious beliefs cause no great harm in society and in a sense, he's correct, because moderate believers can't be held directly responsible for the extreme events we see in the world today. I covered some of the "indirect" responsibilities of moderates earlier in this book. But this brings up an important point; it's always an easy matter to search and find people who agree with you and ignore those who disagree. With that system you can validate almost any claim. But I will remind you again that my goal isn't to simply prove myself right; it's to look at enough evidence, for and against, so in the end we might make better choices in our lives and in the world.

On Dr. Newberg's website you can download research papers dating back to 1992. I might also add that the only thing new about tying religious experiences and neurological disorders together is today's advanced technology, which allows us to peer more deeply into the workings of the brain. As far back as 1892 textbooks were linking what was called "religious emotionalism" and epilepsy together. Nearly a century later, in 1975, neurologist Norman Geschwind of the Boston Veterans Administration Hospital first clinically described a form of epilepsy in which seizures originate as electrical misfirings within the temporal lobes, large sections of the brain that sit over the ears. Epileptics who have this form of the disorder often report intense religious experiences, leading Geschwind and others, such as neuropsychiatrist David Bear of Vanderbilt University, to speculate that localized electrical storms in the brain's temporal lobe might sometimes underlie an obsession with religious or moral issues.

This is from *Scientific American Mind (October 2007):*

> *Such efforts to reveal the neural correlates of the divine—a new discipline with the warring titles "neurotheology" and "spiritual neuroscience"—not only might reconcile religion and science but also might help point to ways of eliciting pleasurable otherworldly feelings in people who do not have them or who cannot summon them at will. Because of the positive effect of such experiences on those who have them, some researchers speculate that the ability to induce them artificially could transform people's lives by making them happier, healthier and better able to concentrate. Ultimately, however, neuroscientists study this question because they want to better understand the neural basis of a phenomenon that plays a central role in the lives of so many. "These experiences have existed since the dawn of humanity. They have been reported across all cultures," Beauregard says. "It is as important to study the neural basis of [religious] experience as it is to investigate the neural basis of emotion, memory or language."*

Dr. Newberg also makes the statement, based on scientific research, that *"Intense prayer and meditation permanently change numerous structures and functions in the brain—altering your values and the way you perceive reality."* Personally I put a lot of emphasis on altered states of consciousness during meditation and the reality of mystical experiences because I have had them myself. However, when the research clearly shows that some of these same experiences can be artificially induced, we must take notice and step lightly into this area no matter how much we want these experiences to be of a divine nature. Used properly, the rigid adherence to the rules of scientific research could lead us deeper into the mystery of consciousness, deeper than we have ever imagined with a mind that was untrained and running wild with untested speculation and superstition. So let's hear what Dr. Andrew Newberg has to say on the subject of beliefs.

Rahasya: I'm doing some research right now on the social and neurological consequences of belief systems and I noticed something interesting in myself while researching. I observed myself getting angry at times when reading about some of the extremists in religious movements. It was almost like there was a mental pattern already there and all I needed was to run some energy or thoughts through it to wake it up. Do you think we may have preexisting neural patterns for fundamental and aggressive beliefs, possibly as a by-product of our early evolution?

Andrew Newberg: Yes, absolutely. My book *Why We Believe What We Believe* does go into some detail about both fundamentalism and how people respond to things that are negative or anti whatever our own beliefs are. I've done a lot of work lately on mirror neurons, neurons that mirror whatever somebody's doing in front of us, so if somebody is being overtly negative, then we will have a tendency to respond in kind. And that is something that is built into the brain. When somebody is speaking in a manner which is opposed to our beliefs, we're also faced with a slightly different conundrum. We use our beliefs to help us survive in the world, and we want our belief systems to work, and we want to feel good about them. If we're confronted with somebody who is trying to shoot down our beliefs it puts our whole approach to existence in jeopardy. And we usually have one of a couple of ways of responding to that. One way is to really reflect on our own beliefs and try to think about why we might be wrong, which, like I said, is really not a very good position for our brains to be, or the flip side of that is to say, "Well, you know what? I must be right; it must be these other people who are wrong." And that sets up a kind of an antagonistic exclusive, hateful kind of relationship because "Why would somebody be telling me my beliefs are wrong when I clearly know that they're right?" And that really goes to show how our brain is built in order to support our beliefs and to essentially reject those beliefs that are going against our way of thinking and our way of trying to support the belief systems that we have that help support our way of living and surviving. So I think that the short answer to your question is that to some degree there is a built-in mechanism by which we respond fairly strongly and fairly negatively to somebody who is being negative or to somebody who is simply disagreeing with us, in which case it's a very unhappy position for our brain to be in. Our brain does not want us to be wrong. Because that has very dire consequences in terms of our overall survival.

Rahasya: Do you think that it's at least possible that as a human race we are entering into a new level of consciousness where we no longer feel the need to invent beliefs when we hit that wall that we inevitably hit when we are searching for answers? Like when you are searching for any truth, there is always a wall that we hit because of the limits of our understanding, and when we hit that wall we have this tendency to make up a belief and pretend we know what's on the other side of the wall or pretend there's no wall there at all. The other option would be to embrace the mystery and be courageous enough to say, "I don't know."

Andrew Newberg: That is true. I think our brain does have a tendency to be true to its own ideas and statements. In my last book about beliefs, I said that everything we do and everything we think about is a belief. Until we get to the point where we look beyond our own ego-self, and to some degree beyond our own mind, we are always going to make assumptions and have beliefs to make our brains feel more comfortable. And if we can get to a point where we embrace that uncertainty and doubt, and be willing to learn from that and to explore that, I think that that could be a very positive experience. And whether or not we are on the brink of evolving into that, I don't know. I certainly know there are people that have been successful or are more successful than others at doing that. But it's hard for our brain to break out of these patterns that have been around for thousands and thousands of years. So it's possible, but it's not easy.

Rahasya: That's true. I stopped believing in a personal God that listens to me many, many years ago. I realize now that it was the mental concept or thought of God that was getting in my way to having a deeper realization. I have friends who are atheists and agnostics and to the surprise of my religious friends they have very deep and meaningful mystical experiences without any concept of God. The reality is that my own conscious experience has gone much deeper after I let go of the concept of a personal God that I was indoctrinated with as a child.

Andrew Newberg: In our new book *How God Changes Your Brain,* I think the answer to your question is that it kind of depends on the individual. The concept of God can be very interfering for some, as in your case, and can be very opening for others. There are many people who say it's not God, or a personal God, but it's an energy, it's a force, it's a unifying conceptualization of the universe. I think for some people it can be a very positive, and a beneficial way of looking at things. But then certainly for others it can get in the way. I think part of the issue that often comes up, which I also think you are alluding to, depends to a large extent on how one defines what God is especially if it becomes exclusive and a hate-filling definition. It may depend a lot on the person's own individual beliefs and past experiences that may have an effect in terms of how a concept of God really does affect them and change them for good or bad.

Rahasya: We really do see the world, not as it is, but as we are. For me, this is even further evidence of the necessity of all of us looking at our beliefs and putting them out there to see if they are valid in this modern world. The fact that most of the world's religious beliefs come from ancient books supposedly written by the creator of the universe thousands of years ago is troubling for me. So you're correct, whether we are good or bad does depend on our concept of God, which is what I see at the heart of the problem in our world today.

Andrew Newberg: A more fundamental question is how do we know what reality is. And how do we get there? That's what I keep looking for.

Rahasya: I think we're getting there with conversations like this and to stop pretending we know things that we have no idea about and then go out to recruit others and build organizations around those beliefs.

Andrew Newberg: The thing that concerns me more than anything is when anybody becomes closed off and exclusive of other people's ideas. All of our brains are in the same mode of searching when trying to understand reality, and when we come to a belief system that makes sense to us we hold it strongly, whether it's an atheist view, or a Christian view, or a Muslim view, whatever, then it's very hard to let go and I think it's so important to try to foster dialogue.

Rahasya: Yes, it gets to be a very tricky situation when you're dealing with religious people that come from an exclusive point of view, where their way is the only way. Then of course when you mix that with nationalism and high technology, which we always do, we have a problem.

Andrew Newberg: Yes, and not just religious. There are plenty of scientists that hold on to theories even in the light of new evidence that contradicts them.

Rahasya: Yes, I talked to Bruce Lipton about that; he spent years being excluded from the academic part of his own world because of his views on the cell, which are just now becoming accepted.

How important do you think meditation is to mental health and dealing with the complexities and varieties of religious experiences we have in our lives?

Andrew Newberg: My research and what a lot of research I think brings to the table is that many of these kinds of practices are very beneficial for people. What we're showing is evidence that one particular kind of meditation, that we've studied at least, has shown improvement in memory, in cognitive function, in mood. A number of, many other studies have shown the effect of meditation on depression symptoms, on anxiety symptoms, so as a general statement, then, and what I think a lot of research ultimately points to, is that many of these practices are ultimately beneficial for people.

Rahasya: This next question is something I'd really like your feedback on, and we might step out on a limb a little bit, to get this, but I think it's really important in today's world. Since we live in a technically advanced society that seems to be holding onto beliefs, some of them that date back to the Iron Age, I see a danger that can no longer be overlooked, especially when we consider a quote by Voltaire when he says, "If they can make you believe in absurdities, they can make

you commit atrocities." The social consequences of maintaining our out-of-date and dysfunctional belief systems are fairly obvious in today's world, but what about the neurological consequences of maintaining those same beliefs? Do you think there may be a neurological connection in the brain of a person who maintains a belief system that is by all definition absurd and the ability of that same person to commit an act that would be considered atrocious by most people? In other words, what is the making of a suicide bomber?

Andrew Newberg: Well, the scary aspect of what you're asking is this: The research that's coming in is clearly showing that we are all capable of such things. If you put people in the right, or I should say the wrong environment, you can get lots of people to do lots of really bad things, even though inwardly or outwardly, they would not generally do that. I certainly don't feel like I would be the type of person that would hurt other people but the evidence points to the fact that I would, given the right circumstances and the right state of mind.

Rahasya: So one might come to the conclusion based on what you just said that there is a possible relationship between a limited and distorted view of reality and your ability to commit atrocious acts. Some of this stems back to a saying in biology, **"Neurons that fire together, wire together."** ***It only stands to reason that if they don't fire together, they don't wire together. So I think that to support a limiting belief you need to cut yourself off from knowledge of science, life experience, and history, which means you actually stop firing those neurons, which is the same network you need for understanding many of the complex situations we run into, and the problem solving necessary to survive and have a meaningful life.***

Andrew Newberg: So that's right, and the plus side of that is the idea that you still need to ultimately focus and concentrate on certain ideas and practices like meditation. And as my late colleague and I always said, these are morally neutral technologies, you can use meditation, and prayer, and ritual to foster compassion, love, and inclusiveness, or you can use them to foster hatred, and exclusiveness, and anger. And it's really just a matter of what concepts, ideas you decide to focus on.

Rahasya: In Chapter 1 of your new book that's coming out, you talk about why religious beliefs generate both anger and compassion in virtually everyone's brain. What is it exactly that determines whether it's going to be anger or compassion?

Andrew Newberg: Well, it is a combination of things. It is a combination of the thoughts that are associated with it and whether the emotions that you ultimately try to foster are going to be positive or negative. We tell the story about a little boy asking his grandfather about that very question, and the grandfather's response is that there are two wolves that are battling it out inside of your mind: One is selfish and hateful and the other is compassionate and forgiving, and the child says, "Which wolf wins?" and the old man says, "Well, it's the wolf that you feed." So, if you continue to feed ideas, concepts, behaviors, and all these things that foster the positive side, then that is the one that wins. Then of course after one side wins, that's the beginning of a belief system that is difficult to break away from.

Rahasya: Yes. It's difficult, because we identify with those beliefs, and to some extent those beliefs become us. We feel threatened; that's the reason we say, "I am a Republican, I am a Christian, I am a Muslim."

Andrew Newberg: Right. In fact, in my talks I give about why we believe what we believe, what I thought was a cute way of ending was to say that we should really take Descartes' old idea of *Cogito, ergo sum* and switch to *Credo, ergo sum,* that "I believe, therefore I am," that it really is those beliefs that make us who we are.

Rahasya: Right. But that's the whole point, that's the place I have arrived at in my life, because I am starting to let go of beliefs and to embrace what's left, which is the mystery. And it can be scary at times when you find yourself face to face with that "nothingness" that seems to be everything the world is made of.

Andrew Newberg: But it's wonderful, and you're able to do it. And I always think to some degree it would be wonderful to have lots of other people feel more comfortable in doing it, and hopefully, we'll get more and more people there.

Rahasya: Yes, because you know what, it might always be a mystery, Andrew. I don't care how far up the ladder you go, whether you're an ascended master or whatever, I have a hunch there will always be this breathtaking awareness that no one really knows the ultimate truth. Isn't that amazing?

Andrew Newberg: Oh yes, absolutely.

CHAPTER 25

Rahasya, Dhara Lemos, Dr. Bruce Lipton, and Margaret Horton

Evolving toward Consciousness An Interview with Dr. Bruce Lipton

This is one of many talks Dr. Bruce Lipton and I have had on the subject of beliefs and conscious evolution. Dr. Lipton is an internationally recognized authority in bridging science and spirit. He has been a guest speaker on dozens of TV and radio shows, and keynote presenter and speaker for national conferences. We all know the connection between the mind and body is the key to health and here is the science that proves how and why holistic health therapies work!

Recent advances in cellular science are heralding an important evolutionary turning point. For almost 50 years we have held the illusion that our health and fate were preprogrammed in our genes, a concept referred to as genetic determinacy. Though mass consciousness is imbued with the belief that the character of one's life is genetically predetermined, a radically new understanding is unfolding at the leading edge of science and will eventually filter down through society.

For instance, in a quantum experiment commonly referred to as the double-slit experiment, a stream of electrons will present itself as a wave. As soon as the particle is observed by a scientist or researcher, what was once an electron appearing as a wave will disappear and collapse into a particle. Furthermore, it was found that the location where the particle is seen can be determined probabilistically. This was an astounding discovery almost a century ago, because before these discoveries it was generally thought that the nature of the physical world could be determined through classical physics and was independent of the observer. This probabilistic nature of the position of the electron was termed "indeterminacy." We are deeply and consciously connected with the universe we live in and *the universe that lives in us.*

In *Power vs. Force*, Dr. David Hawkins writes: *"In this interconnected universe, every improvement we make in our private world improves the world at large for everyone. We all float on the collective level of consciousness of mankind, so that any increment we add comes back to us. We all add to our common buoyancy by our efforts to benefit life. It is a scientific fact that what is good for you is good for me."*[62] Dr. Hawkins has backed up his remarks and conclusions with 29 years of hard research.

More recently, Dean Ornish, MD, in the article "Changing Your Lifestyle Can Change Your Genes" for *Newsweek* magazine (June 17, 2008), writes, *"New research shows that improved diet, meditation and other non-medical interventions can actually 'turn off' the disease-promoting process in men with prostate cancer."* He goes on to say, "*Here's some very good news: your genes are not your destiny. Earlier this week, my colleagues and I published the first study showing that improved nutrition, stress management techniques, walking, and psychosocial support actually changed the expression of over 500 genes in men with early-stage prostate cancer. This study was conducted at the non-profit*

---

[62] Dr. David Hawkins, author of *Power vs. Force,* 2002, uses kinesiology extensively in his research and I see his work to be a step in the right direction, but we need more research to validate everything he claims.

Preventive Medicine Research Institute and the University of California, San Francisco in collaboration with Dr. Peter Carroll, Dr. Mark Magbanua, Dr. Chris Haqq, and others."

Cellular biologists now recognize that the environment and more important, *our perception of the environment,* directly control the activity of our genes. So as you read and comprehend what is being said here, be prepared to be inspired and at the same time, have your mind engaged in a way that only hard science combined with a passionate spirit can accomplish.

Rahasya: What is the importance of looking at our past and what we believed to be true?

Bruce Lipton: The new evolution that the new biology is suggesting is one that is repeating patterns of self-similarity; it's a fractal.[63] And the significance of the fractal is that if you can understand the previous half of what happened at that time, you can apply the character to what's happening at this time.

Rahasya: Like the Fibonacci patterns found in nature?[64]

Bruce Lipton: Right, so basically what it says is this. The evolution of human civilization is like that of a giant organism, where we are the cells in that organism. And its relevance is that civilization evolved in the same way as an animal; we go through evolutionary phases, which are redundant to previous evolution patterns. For example, for the vertebrate animals the pattern was fish, amphibians, reptiles, birds, and mammals, and there was an evolutionary jump each time. We're in a jump now between a reptilian-based civilization and a mammalian-based civilization, and we're currently headed by reptiles. And yet mammals such as yourself and myself are in the process of rewriting a way of

---

63 Fractal: A geometric pattern that is repeated at ever-smaller scales to produce irregular shapes and surfaces that cannot be represented by classical geometry. Fractals are used especially in computer modeling of irregular patterns and structures in nature.

64 Fibonacci sequence: The sequence of numbers 1, 1, 2, 3, 5, 8, 13 . . . , in which each successive number is equal to the sum of the two preceding numbers.

civilization that's different from the one we have now. The reptiles are conscious but not self-conscious. What that means is that they will live for the moment but they don't have any vision about what the next week is going to be.

Rahasya: It would be nice, if not imperative, to stop repeating reptilian patterns as we evolve into a more advanced civilization.

Bruce Lipton: A perfect example is going to war in Iraq with no plans of what to do six weeks later. That is an excellent example of reptilian thinking. The mammalian civilization is a very different one because the character of mammals is nurturing, and the fact of nurturing means nurturing the planet, and right now we're like dinosaurs raping the planet.

We mammals started this a long time ago, and in fact, I think it was actually seeded in 1969 when we were in the bird phase—the bird phase was Wilbur and Orville Wright getting off the ground until its fullest evolution, when we landed on the moon, and we took a picture of the earth. When we took a picture of the earth in 1969 and it was sent back here, it changed the mind-set for a lot of people: "Oh, my God, that's all there is. We have got to take care of that." And that's when all of a sudden we got the idea to take care of the air, the water, and all of that came from political decisions. And that's because there were enough people who were mammalian in character saying that we must take care of those things for our future so they'll be there.

And so we're working on this, and it's been an uphill battle, and it's coming to a point where the dinosaurs (the reptilian system) is going to crack. In the past, the dinosaurs went through waves of extinction, something happened, and we're about to hit something that will turn under the current system so it can't survive.

And then a new system, a mammalian-based one, will come out of the ashes of that, but you can't have a mammalian-based system until the other one crashes, so that means you have to sit here and keep preparing for the future and also recognizing while you're doing this that all around you things are going to look pretty chaotic.

Rahasya: As you probably know, in most chaotic systems there are hidden patterns that can be seen if you're not consumed by fear.

Bruce Lipton: Well, that's exactly what it is and that's where the fractal geometry comes in. Because the fractals are based on this chaos and order comes out of the chaos and we'll go from the reptilian structure through chaos and then evolve to a mammalian structure.

Rahasya: As I look at some of the sciences, like physics for instance, I've noticed a trend to view life, and our existence in it, as a more cooperative relationship. I also see in your view of cellular life this same tendency. Do you think this is an indication that we as human beings are evolving into a higher level of human consciousness?

Bruce Lipton: The answer is absolutely yes and it deals with the parallels that occur as to why we got here. For the first 3 billion years of life on this planet, all there were were single cells. While they appeared to be like individual living entities, over time they actually learned how to communicate across space with each other so even individual bacteria are in fact, part of a community. And the community is based on the fact that they exchange information with each other through chemicals, some sending out chemicals and other ones responding. These chemicals reflect the status of life or information so others in the community can get information or know what's going on in the community. And the reason why we have intelligence is because evolution is based on an expanding awareness. Some people like to say there is no direction to evolution, but indeed there really is. Evolution is an explosion like the Big Bang. It starts at a certain point but it doesn't go in one direction, it's actually like a sphere, and what it does is it fills space. And in evolution cells fill the spaces.

And while there was a lateral progression, to the left and to the right, we didn't really elevate. But at some point when you're building left and right, then the next step is to elevate. One of the characteristics that's uniform throughout the entire spectrum of evolution is the more complex an organism is, the more awareness there is.

So from the evolution of prokaryote to eukaryote, one of the obvious things to look at is the increase in membrane; there's 100,000 times more membrane in a eukariotic cell than there is in a bacterium. Remember, it's not the genes that control but it's the cell membranes. The only membrane in a bacterium is just the outer skin. When you look at the higher-level cells, you've got all those things: organelles, mitochondria, nucleus, Golgi, endoplasmic reticulum, things like that. All those are membrane derivative. And so what it means is more membrane, more awareness. But then there's a point where you can't make the membrane any bigger because then you're dealing with the physical pressure of fluid inside a bag. There's a certain amount of fluid pressure on the membrane—if it gets too big, the bag can't hold it, sort of like a water balloon. And so there's a point in realism that says that to make a viable functional cell, it'll reach a certain size and if it gets any bigger than that, you can't hold the membrane together. It'll break and then the cell's dead. So that's why evolution went from prokaryote to eukaryote. Three billion years. You know—maximize the fullest evolution of a single organism because the most awareness in a single cell gives us the most evolutionary advantage. But then we reach that parameter that says if I add any more membrane now, I jeopardize my viability because it could rupture. And so it ended; it said, "That's it. I can't make it any smarter and stay viable."

So evolution, in our perception of it, seemed to have flattened out—it's got all these single cells. The next level of evolution was, "Well, if I have, as a cell, X amount of awareness, and you are a cell, with X amount of awareness, and you and I can plug in together, and we can share directly our awareness at the same time, then we have collectively two X's more awareness." And the relevance of that is the efficiency of cells coming together and coming into community and leading to higher levels of awareness. But a community of cells is recognized by the shape the cells take. So one group of cells is called amoeba, a single-celled amoeba, and a group of them might be called a hydra. Another, a larger group, might be a clam. A bigger group might be a cat. And a larger group of them might be a human. It's a community, and looks like an organism, and we give the name of the organism to the shape of the community. And so in our bodies are 50 trillion cells communicating. For what? Advanced evolution. Over the millions and millions of years since the origin of multicellullar organisms, the destination was to make the smartest organism.

Well, then, the smartest cells join together to make a bigger organization and then you get the smartest organization, which is us in a sense, but it may not be—because, it is probably dolphins—but the significance of this is we've maxed our evolution. It is like, "OK, you've maxed out; you're given the exact brain that you have within a living organism, and it's maximum efficiency of awareness." Maximum. What does that mean? Well, then it seems like, "Okay then, that's the end of human evolution." And the fact is, that when we had single cells, what did they do? They joined up in a community to share their evolution. We have single humans. What do they do? Well, they join up with other humans to do what? To share their awareness. So the point is, we are creating a higher level where we are the cells in a larger organism called humanity.

Rahasya: It's interesting how we are manifesting that in our outer world as libraries, postal services, and now the Internet.

Bruce Lipton: Oh, the Internet especially, because that's what cells mostly did; they were able to instantly connect to the same information at the same time so that a skin cell, a liver cell, and a bone cell instantaneously know when you're having a bad moment.

What this means is that we're going to have 6 billion different perceptions giving us a full spectrum of awareness. And so as much as people are fighting it, the reality is an inevitable evolutionary step in which we're going to leave behind our reptilian individual fight-or-flight behavior, you know, the struggle for existence kind of stuff—Darwin. And the next level of evolution is to recognize that it is a community that creates.

Rahasya: And it seems like it takes so many subatomic particles to make an atom, so many atoms to make a molecule, so many molecules to make a cell, and it might take so many people to bring this about.

Bruce Lipton: Well, I think so, because if you look at it, each human is like another eye on the universe. And the more eyes, the more awareness. It's simple logic. That's why this is going to be a very quick evolution, because it's a matter of wiring. The faster the wiring's in place, the faster an idea runs through the system and changes things. So with all the wiring in place, how long does it take a piece of news to affect the world? When the wiring's in place, what happened 10 minutes ago is known by everybody on the planet.

Rahasya: How do you feel about society's trend to self-medicate for stress with drugs such as Prozac and Zoloft?

Bruce Lipton: Well, how I feel about that is it is yet another example of a consequence of educating people that they are victims of their machinery. And as a result, we don't take responsibility for the experiences we have in our bodies. We blame that on the mechanism that is failing. So, for example, we can talk about cancer and say, "What's wrong?" "Oh, the mechanism is defective and that's why it happened." And the reality is, less than 5 percent of cancer is hereditary in the first place; 95 percent of cancer is due to how a person lives his or her life.

It's the same situation with cardiovascular disease. When you have a heart attack, does that mean you had a bad heart? No, it just means you were a lousy driver and you stressed the hell out of the system. Taking a drug is tantamount to taking your car and putting masking tape over all the gauges because you don't want to see what the gauges say. But the problem is, somewhere down the road you're going to have a steaming burned-out mess.

And the reality is that anyone who takes a drug to cover up the symptoms never really gets to what the cause of the symptom is; they just took away the expression of the symptom.

Rahasya: I'm sure you remember Linus Pauling; he once said this, **"Everyone should know that most cancer research is largely a fraud and that the major cancer research organizations are derelict in their duties to the people who support them."** ***Since the mid-1950s there has been well over 1 trillion dollars spent on cancer research and the American Medical Association still only recognizes three ways to deal with it: surgery, radiation, and chemotherapy.***

This is good information for people to really start reevaluating their lives and their actions and responsibilities. In the movie **What the**

Bleep Do We Know, ***they said that most of our bad decisions are simply because we have bad information, which leads to limiting beliefs that we have about ourselves and reality, most of which is unconscious.***

Bruce Lipton: Yes, the way we live our lives, we run most of it on unconscious tapes and our minds are in the future or the past. The dangerous thing is that the tape is being programmed by other peoples' behavior and not even known to us in our consciousness. We have no idea how many of these limiting programs exist. Even if your beliefs are in your conscious mind, it still never changes the program in your subconscious mind. That's why people have to use so much willpower to override the subconscious mind and that's where the problem of power comes in, because the subconscious mind is, I think the number is 5 million times more powerful than the conscious one. There are 20 million bits of information processed every second in the subconscious mind and only 40 bits of information are processed in the conscious mind every second.

Rahasya: Which shows you the power of the filtering process created by our beliefs.

Bruce Lipton: Absolutely. It runs the show; psychologists say about 85 percent of the day is run by the programming of your subconscious mind, most of which are limiting programs. Therefore, 85 percent of our life is trying to go uphill against ourselves.

Rahasya: It seems like the quest and the thrust of life is to self-organize into higher and more complex organisms for the soul purpose, that's soul, s-o-u-l, of evolving into ever-higher levels of awareness and consciousness. Do you see us continuing this process as human beings linking together to form a planetary consciousness?

Bruce Lipton: Well, of course! Gaia, the planet, is a living cell. We are the membrane, we're the proteins and the membrane, and if you remember, the proteins and the membrane are the awareness units. We are like antennas downloading information. And the beautiful part about that is when we wake up and find that we are all in the same organism, that completes the earth as a single living cell. Then, as in all the evolution that led to this point, this single cell (Earth) then joins up with the other cells to share their awareness. And so on and on . . .

Rahasya: Then we're talking about solar, then galactic consciousness that's so far beyond even our wildest imagination.

Bruce Lipton: Absolutely. And we have hints that it's there, but we're not! And therefore, we're not able to be in that dialogue with "them" yet. When the earth completes its process, and when the earth is a single mature cell, at that time it will hook up with other mature cells, other planets, with life forms. It's a fractal thing . . . it's implied that it must go that way.

Rahasya: It's an exciting time to be alive, isn't it, Bruce?

Bruce Lipton: It's exciting because we see the possibilities for the future, and we're living in the light. As for the other people that are locked in their fear of death and the struggle, when it comes down to it, one side's going to walk away from this and be whole and the other side's going to have to die because that's their vision. And I like it that we're on the side of the light.

Rahasya: So do I, Bruce, so do I, and you know what? I think it's really great that you're putting out information like this. Validation somehow expands our awareness, which in turn gives us the courage to evolve past outdated beliefs and in the process help some of those who are stuck to have a different vision of what's possible.

Bruce Lipton: Likewise, what you're doing in your community adds to this awareness; to get every cell to turn its light on is what creates evolution, so we're all doing this together.

For more information about Dr. Bruce Lipton, visit www.brucelipton.com.

CHAPTER 26

Rollin McCraty at HeartMath

The Coherence of Heart and Mind An Interview with Rollin McCraty from HeartMath [65] [66]

[65] Rollin McCraty, PhD, is the executive vice president and director of research at the Institute of HeartMath, and he has been with the organization since its inception in 1991. He worked with founder Doc Childre to formulate the research goals of the organization and create its scientific advisory board. Dr. McCraty is a fellow of the American Institute of Stress, a visiting senior scholar at Claremont Graduate University, and a visiting professor at University of Alabama at Birmingham. He has coauthored and published numerous research papers.

[66] HeartMath LLC is a cutting-edge performance company providing a range of unique services, products, and technology to boost performance, productivity,

My interview with Rollin McCraty was exceptionally interesting because he brings to the table a vast repository of documented research on the interaction between the heart and the brain. Rollin comes from a very traditional scientific background and is one of the many scientists today who are stepping beyond traditional paradigms. This is not to say that he is stepping beyond the rigid scientific controls for research. The research at HeartMath has shown the importance of heart and brain coherence and the possibility of our future evolution toward higher awareness through this coherence.

Rahasya: Something that got me started down this avenue of writing a book on beliefs was when I read a quote from Voltaire that said,* "If they can get you to believe in absurdities, they can get you to commit atrocities." *And it started me thinking, "Why is this so?" And the deeper I get into it the more I'm convinced that the imprint or neural network that's associated with believing something absurd is the same neural network or imprint that a person has and is easily manipulated into doing things they would normally never do, like going to fight in a war that is not justified. In other words, if I can get someone to believe in something totally absurd, disconnected from reality, illogical, and even counterintuitive, which a lot of our beliefs that come from the Stone Age are, it seems like it's the same type of cognitive processing that it takes to actually be able to manipulate that person into doing atrocities that normally that person would never do. How do you feel about that?

Rollin McCraty: I know what you're talking about. And it's certainly true. It's a complex situation, because I think there are a number of factors that come to bear on this. One of the exercises we do in the HeartMath training is what's called a heart map and a mind map on a subject. So you have people really think out and do all the things around an issue or a challenge they might have, and then we have them get into coherence between their heart and brain. Then they ask the same question of

---

health, and well-being while dramatically reducing stress. Founded in 1991 as a nonprofit research organization by Doc Childre, HeartMath has earned global recognition for its unique research-based techniques and proprietary technology to transform the stress of change and uncertainty. www.heartmath.org.

themselves, ask their heart, ask their intuition for some perspective on it, and then we list all the things that come up after they do that exercise while in coherence. And the difference is striking with those two perspectives. And almost invariably there's a profound difference, and then we ask people, "Okay, just reflect on these two perspectives a little bit." And one of the things that becomes very obvious is that the more cognitive approach to it is far less rational than when they ask their heart for the emotional perspective on it. It's a much broader, more inclusive, and sometimes just common-sense kind of approach that they come up with. So they start saying, "Wow." This heart intelligence, as we call it, the more intuitive, spiritual intelligence, is far more rational than what they got from the mind. So I think ultimately what this is, in the context of what you're talking about, is really disassociation. If you can get people to dissociate from their intuition, from their core values, and come more from the cognitive mind, it's exactly then when we are able to do those things that we would normally never consider. And it's that disassociation, or separation, that the human mind is so good at doing.

Rahasya: Yes. It seems like it's the human mind, the actual brain that can absorb things like fear, anger, jealousy, and defense and so on.

Rollin McCraty: That's all the mind stuff. It's the separation, and the ability to disconnect from ourselves and from the other people. The cognitive domain of the mind is really a different intelligence than the domain of the heart is. I mean that literally, by the way. So all of the atrocities and the crazy things that we do are done by people who are stuck in a mind-, or brain-dominated consciousness.

Rahasya: Yes. I know for a fact, Rollin, it would be impossible if I'm in my heart to do any damage to any person on this planet.

Rollin McCraty: Exactly. That's really my point. And it's that disassociation between the two. Obviously we need our brains and our mind, that's the equipment we need to live in a three-dimensional world, the sensory intake and labeling and all the processing it does, so it's not about good-bad or one or the other, it's about the integration of heart and brain.

Rahasya: I think it's a matter too of whether your brain is a master or a servant.

Rollin McCraty: Well said. You could be a HeartMath trainer. The cognitive and mind-brain systems becoming the servant of the heart and our heart intelligence, which is really our spirit, our spiritual, our higher-self intelligence that we're talking about.

Rahasya: I think the human race has vistas and visions and states of conscious evolution that we can barely even comprehend or even imagine with our mind.

Rollin McCraty: Absolutely. That's where intuition comes from; it's really our higher self, our spirit self, which is part of our inner being.

Rahasya: I would like to get your opinion on something else in the area of beliefs. Since you have done so much research in the area of the brain and the heart I was wondering if you have any ideas on how a person can be so blinded by their beliefs?

Rollin McCraty: Are you familiar with "attentional blindness"?

Rahasya: No.

Rollin McCraty: It is relevant to your question. First, it's not a concept, it's a scientific fact. Attentional blindness.[67] In other words, it's a way of showing that we literally only perceive what we pay attention to.

---

[67] *Attentional blindness,* also known as *inattentional blindness,* also known as *perceptual blindness,* is the phenomenon of not being able to see things that are actually there. This can be a result of having no internal frame of reference to perceive the unseen objects, or it can be the result of the mental focus or attention that causes mental distractions. The phenomenon is due to how our minds see and process information. Humans have a limited capacity for attention, which thus limits the amount of information processed at any particular time. Any otherwise salient feature within the visual field will not be observed if not processed by attention. Also related to this is the phenomena of blind people who later in life gain sight. Their processing of the visual stimuli

We did a lot of research on this, and there's this little film we made to illustrate this. We have six people running around with two basketballs, and we have two teams, one with white shirts on and one with black shirts on, and they're passing the basketball around. And, so you've got two balls in play, one to the black team and one to the white team. So they have the ball and they're passing it around. Sometimes they'll bounce it, then pass it and sometimes they'll throw it through the air; sometimes they'll just go around in circles, you know, just playing catch. So the instructions for watching this video are, okay, you are to count how many times the people with the white shirts bounce pass the basketball. We emphasize bounce pass, not air pass, make sure they understand the instructions, and kind of get a little competition going if you can, depends on how many you've got in the room, so you play this film. And then it's amazing how many answers you get. But that's not the point. What happens during this, is in this case, a gorilla walks in the middle of them while they're passing the ball around; it's a guy in a gorilla suit, obviously stops in the middle, stops right in the center of them, beats on his chest, jumps up and down, and walks on through. Afterwards, you ask the participants who were watching the film, what's the count? And a few of them get it right and a few of them get it wrong and everybody laughs. And then you ask if they've seen anything unusual. And almost always, they will say no. They'll talk about the pattern, they were going in circles, or this or that, and then you say, "Okay. I'm going to show you the film again. This time don't count anything."

They all see the gorilla. So what we did in that exercise is we focused their attention on counting. They completely missed the gorilla.

Rahasya: So if your attention was focused on defending your beliefs it might filter out important information to the contrary.

Rollin McCraty: Yes, that's the whole point; in other words, if you don't attend to it, you don't even perceive it. Now, we actually made this film for one of our programs for kids called Test Edge. It shows them that

---

does not allow them to identify objects easily; effectively they can see but are still perceptually blind. From Wikipedia

when you're not paying attention, you don't perceive, you literally don't perceive it, and this is exactly what happens in high-test-anxiety kids. So if you're in a highly anxious state, where's your attention? It's split. So these are the kids, at best, who literally don't see all the words or sometimes entire sentences in a test question.

Rahasya: So anxiety could be the constant bombardment of news that seems to have the purpose of creating fear and anxiety. I remember counseling adolescents going through drug rehab and when they would come in to the program they were so full of anxiety and defensive that there were some things that they simply didn't hear.

Rollin McCraty: No, they heard it. We've proved this in studies that we've actually done here in our own lab and others have shown this also. In other words, if you're doing neural recordings, you'll find that in cases, whether it's visual or auditory, the brain got it. All the sensory systems worked, but it gets blocked and we just don't perceive it but the signals are there.

Rahasya: So how can we unlock that?

Rollin McCraty: By paying attention to other things and not being so rigid in our thinking and our beliefs. Now, I also use this example in our work on intuition or the signals from the heart. Because we're literally suggesting that in this coherent state, that that helps align, certainly the emotional and cognitive centers but also the connections to which most people commonly refer to as the spirit or higher self, which is certainly in our models. The main pathway through that is literally the human heart. So if intuition uses the emotional body for those signals to filter through and we don't attend to them, if we don't pay attention to them, we don't perceive them.

Rahasya: So if we don't bring it into focus through our consciousness it simply isn't part of our conscious reality.

Rollin McCraty: Right. The signals are there and our intuitive signals are going on all the time; we just don't pay attention to them and thus we don't perceive them.

Rahasya: That's what I find most surprising of all, sometimes when I listen to someone talk about their beliefs, it becomes clear that what they believe is not only irrational but also counterintuitive. All of this cuts down on our ability to make appropriate decisions in our lives.

Rollin McCraty: Exactly. That's what intuition is really all about in the way we're teaching it and helping people get in touch with it. So getting in this coherent state, we're increasing our access to our intuition. So intuition is really the information that our higher self is trying to tell us in our day-to-day decisions; it's not just about trying to see through a deck of cards or trying to win the lottery or inventing a new light bulb. Yes, that's an aspect of intuition, but the real gain, the real benefit, is in our day-to-day lives, in the discriminations, in the decisions that we make.

Rahasya: How do you feel about synchronicities?

Rollin McCraty: They're great; that's one of the things that we've heard for years and hundreds, thousands of times, probably, of people who start learning the HeartMath tools and how to get in this coherent state is that the synchronicities in life, just that they can note, take a jump that's beyond what they can write off to as coincidences.

Rahasya: Yes. Because the more I pay attention to synchronicities, the more they seem to happen. Of course they may have been there all along and by putting my attention on them I notice them more; either way they become part of my conscious reality.

Rollin McCraty: Yes. It seems as though those things are organized more at the spirit level and I believe that is the only real explanation for a lot of what we're talking about here.

Rahasya: It's inspiring and gives me hope when I talk with people like you that are coming from the traditional and conventional point of view, but once you start expanding and you notice that your consciousness is expanding beyond what tradition can explain, you start stepping into this exploration of consciousness itself.

Rollin McCraty: Right. And in fact, the HeartMath tools are for rapid growth in consciousness.

Rahasya: And what could be more important in today's world?

CHAPTER 27

Rahasya and Byron Katie

The Ultimate Religion An Interview with Byron Katie[68]

I had the pleasure of spending some time with Byron Katie and of course I wanted to know how she felt about my view on beliefs. When it came down to it, she agreed on many points except one. Even though we agree that all of us need to do our own work on ourselves, we disagree on how much someone else can play in that process. It seems to

[68] Learn more about Byron Katie by visiting www.byronkatie.com.

me that there are a lot of people on this planet who simply need better and more accurate information and then they will start looking into their beliefs. But Katie holds the opinion that to take a person's beliefs away prematurely is not right and in that sense she is correct. You could even see that process as being aggressive, even if it is only on an emotional level. It is for this reason that I have edited this book several times so as not to come across as aggressive or angry. However, I'm not perfect and Katie, being the person she is, was quick to point that out to me and I'm a better person for it. Life is a constant learning lesson and it's truly a journey, not a destination.

Rahasya: Tell me a little bit about your work, Katie.

Byron Katie: Okay, The Work is four questions and a turnaround. The Work is a way to identify and question the thoughts that cause all the suffering and violence in the world. The Work is a very, very simple process. It's for anyone who can answer a question that is willing to. It takes a bit of willingness and an open mind.

Rahasya: So, what are the questions?

Byron Katie: The first question is: Is it true? For example, if you had the thought that he doesn't care about me, you would simply write that thought down, the thought that you're assuming is true. And then you would question it, Is it true? Second question: Do you absolutely know that it's true?—that he doesn't care about you. Third question: What happens to your life, how do you react when you believe that thought? Or, as I like to say, how do you react when you think that thought? What happens?

Rahasya: It seems that you are talking about an actual physical feeling, not just an emotion.

Byron Katie: Yes, just close your eyes and feel how you react when you think that thought. And then you begin to notice the feeling. I can't stress this strongly enough, just see the reaction that happens. Also notice the images in your head that happen and out of those feelings and images, how you treat yourself. And then how you react toward others and your life and your business world, your family life, your private world; these are all reactions to the world that are created out of believing just this one simple concept or thought. The world begins with what we believe. And then the fourth question: Who would you be without that thought?

Rahasya: In other words, how much are we identified with it?

Byron Katie: Well, actually, to sit still in the question, to notice all those images, just to see yourself in the marketplace, to see your life and just *notice* what it would be like without believing that thought—he doesn't care about me.

Rahasya: It seems like you keep catching me because my mind keeps kicking in. You're talking more about being a silent witness to this, just observing.

Byron Katie: Yes, because it's meditation.

Rahasya: So we are in a process of putting the light of consciousness on it.

Byron Katie: Absolutely. That's very well said . . . putting the light of consciousness on it.

Rahasya: And then what happens, Katie?

Byron Katie: And then I like people to turn the concept around and ask what it would be like if it were the opposite; for example, if I have the thought that he doesn't care about me, what would it be like if he did care about me? And the mind does not like that, because it's contradicting what it believes and believing is how it stays identified with what it is—which is who you believe yourself to be. Just notice the feelings that happen and the resistance there. And just notice and continue to find examples that would lead you to believe that he actually does care. And then look at another opposite, which might be, I don't care about him. And this can be very shocking. Then you might see that you have been just as unkind or even more unkind than he was.

It's really an eye-opener. So, that's how to work the turnaround; finding the opposite examples. So four questions and a turnaround. I like to say The Work is: Judge your neighbor, write it down, ask four questions, and turn it around.

Rahasya: Yes, that would allow you to see different things. I know there've been times in my past, one relationship in particular—I swore up and down she didn't like me, so obviously the relationship was not going to work. Later on in retrospect, with the advantage of hindsight, I realized that it wasn't her who disliked me, I didn't like her and I didn't want to take responsibility for it.

Byron Katie: Oh, my goodness, yes. And that's what The Work is about. Just because we love someone, we begin to notice that love is an experience and we have to do something with it. We don't have to marry people or date them because we care about them very deeply. It's an experience.

Rahasya: I'm curious about something, because as you know, I'm writing a book on beliefs. I really see that there's a possibility that this insatiable desire we have for believing, which is nothing more than pretending that you know something that you actually have no way of knowing, may be a primitive cognitive process and we're evolving into an area where we are more spiritually comfortable living in the mystery, with less fear, so we no longer need to make up stories to pretend we know something about a seemingly hostile world.

Byron Katie: We are always comfortable living in the mystery because there is no future. It's simply The Work allows us to be, not only comfortable in the mystery, but to realize that the universe is friendly, and it is so friendly, in fact, that we don't need to plan, we don't need to take care, that it's all within us, it's all being done. And anything that I personally would plan would shortchange myself. So I don't bother. I'd like to say if I needed a future, I would get one. But it's not true yet.

Rahasya: But doesn't your ego need that?

Byron Katie: Yes, that's the standard ego, and that's for sure.

Rahasya: What if a person came to you who was really stuck in a religious belief and they really had this belief that the other person, the other religion, the other belief, was out to get them? How would you start getting them to see their own beliefs?

Byron Katie: Oh, I absolutely would not. I would just work with my belief about them. We can't take another person's belief from them. That's their whole world. That's like trying to steal the world from someone. The most we can do is work on our own beliefs. And we become so attractive in that genuineness that "others" dare to drop the religions that frustrate and divide them and begin to feel safe with someone who is loving and kind and caring; that's the *ultimate religion* and we all know it.

Rahasya: So, by you working your beliefs, how is that going to help them, like in a workshop situation?

Byron Katie: Well, I'll be working with someone who is open to it, and then the audience's beliefs fall away. They just sit in the presence of this. You know, sitting in the presence of truth. So the person that I am working with, the person that is answering the questions, that is the presence of truth. And it changes all of us.

Rahasya: Yes, I guess that's where you have to trust.

Byron Katie: Well, whether we trust or not, that's how it is.

Rahasya: That sounds good to me because that removes the need to have faith, which is the fuel for most beliefs. You've seen your Work enough to where you're confident in that it . . . works?

Byron Katie: Yes, whether we trust or not, it is what it is and it works, and we work on ourselves. That's as good as it gets.

Rahasya: So can you tell me one instance where someone's come to you and you've seen the most dramatic effects?

Byron Katie: Oh, the most dramatic effects. I've watched schizophrenics come to love themselves. I've watched multiple personalities and bipolars do the same, but the most dramatic one that I can really testify to is my own. And I don't call it The Work for nothing.

Rahasya: Because it is work.

Byron Katie: It's work. When you are level, you do The Work. And then as your moods begin to swing, you are willing to look forward to doing The Work because it shows you what you are assuming, again, that would bring on that stress, that anger, that rage, that sadness, that frustration, whatever it is. If it's not love, I would question what I am assuming. And that's it. That's the key. To question what you are believing.

Rahasya: It seems to me, Katie, that this would work with people who are having a difficult time with religious belief systems, especially ones that are divisive and limiting, which most are.

Byron Katie: Oh, it absolutely works but everyone's mind isn't open to it.

Rahasya: I live in a pretty progressive community and I see so many people questioning their beliefs, and they're going back and reading the Old Testament, and they're saying, "Wait a minute, I don't really believe this."

Byron Katie: Marvelous. They're getting their own minds back.

Rahasya: Yes. My dad lives with me, he's 83, and he came to me the other day, and he said, "Son, look at this," and I read it, and it said if your son drinks too much, or he's unruly, you're supposed to take him to the edge of town and stone him to death. And I said, "Yeah, Dad, there's a lot of stuff in there like that." And I can see it go through him, at 83, questioning his beliefs, and it's really difficult. And this would be a good little system to do.

Byron Katie: Oh, honey, it is so true, and it's nonthreatening. It doesn't say you have to answer the questions, there are no rules, no regulations.

It's just an opportunity to take your own mind back and begin again, not with a rote belief and not what you've been told, but what do *you* think?

Rahasya: What turned him around was the feeling that it evoked when he had the thought that he would need to kill me because of a belief that his system supported.

Byron Katie: Yes, Exactly. How he reacted when he read that part was his turnaround.

Rahasya: It almost seems like the system, what you call "The Work," is a natural thing that we do.

Byron Katie: It's absolutely natural; it's watching how you react when you believe a certain thought. Is it true? It's in the Bible, yes. And do you absolutely know that's true? Yes. It's in the Bible. How do you react when you think that thought? So who would you be without the thought? And then he gets it. That's an easy one to get. And then a turnaround, "If my son drinks, I should stone him to death," then turn that around, "If my son drinks, I should not stone him to death." So for your dad it's a matter of asking which one of those are truer? And it's absolutely natural. And when a person works with these questions for a while, along with the turnaround, it becomes wordless, and it's alive, and it never ends, and your mind is free, and it's open.

Rahasya: Yes, I saw him going through an inner transformation in which he felt empowered too. Because he took responsibility for questioning, and it took a lot for him to question that at 83 years old because according to his belief system he was risking eternal damnation for even questioning the Bible. I think I was more proud of him in that moment than any other.

What's so intriguing about The Work that you're talking about is this, ever since the first time I talked to you, I've been working on myself, because as I'm writing this book, it pulls up feelings of anger. And I don't know if it's past life-related, I really don't care, and I don't think it even matters. This has really helped me to write, embracing

compassion, because if we keep doing it the way we've been doing it, it's just going to get worse.

Byron Katie: Yes. Oh, good for you. Good for you. And I really want to support that.

CHAPTER 28

Shakti Gawain

Visualizing a New World
An Interview with Shakti Gawain[69]

I was curious how we might look at the conflict in the world as a reflection of what is actually inside ourselves and what better person to ask than Shakti Gawain, who wrote *Creative Visualization* and believes that we mirror our world in our relationships. I also wanted to get her feedback on meditation and, in particular, active meditation, because it has been shown by very carefully documented research that you can predictably reduce the crime rate in a major city by getting a certain number of people to meditate. If that is the case, then we all have the power to help change the world simply by meditating and changing ourselves.

Shakti has been a pioneer in helping us visualize a better world and taking responsibility for our lives. Shakti is also a passionate environmentalist and believes that by bringing awareness into our lives we can bring balance on our planet. This is one of the best examples of using the power of belief or intention the way it should be used.

---

69 Shakti Gawain is a pioneer in the field of personal development. For more than 25 years, she has been a best-selling author and internationally renowned teacher of consciousness. Shakti has helped thousands of individuals in developing greater awareness, balance, and wholeness in their lives. Shakti has written numerous books considered classics in her field. Her distinguished publishing history includes best-sellers *Creative Visualization* (1978, 1995, 2002), *Living in the Light* (1986, 1998), *The Path of Transformation* (1993, 2000), *Four Levels of Healing* (1997), *Creating True Prosperity* (1997), and *Developing Intuition* (2002). Her books have sold more than 10 million copies and have been translated into more than 30 languages. She is the cofounder, with Marc Allen, of New World Library Publishing Company. For more information visit www.shaktigawain.com.

Rahasya: We forget that life is a process . . . . Do you think the trend that we see in the world today to unite and come together (and also the conflict in doing so) is a reflection of what's happening on an individual level within ourselves?

Shakti Gawain: Yes, I do. Absolutely. I think that everything in the world around us on one level is a reflection of what is going on inside of us. So each of us as an individual creates a life—we draw to us certain people and events and circumstances that reflect what's going on inside of us, so we can literally look at our life and see a mirror of our own consciousness. And if that's true on an individual level, it's also true that what's going on in the world in a bigger way is a reflection of the collective consciousness. All of us as human beings are connected by one energy and consciousness. So we all affect each other and everything that's happening in the world affects us individually. And what we're each doing as an individual affects what's going on in the world.

It's a wonderful mirroring process that we set up to help us learn, to help us grow. So anything that's going on, for example, a person in my life that's important to me is always a reflection of something that's going on inside of me. It's interesting to do that on a personal level—to start to look at how the people in your life reflect the different parts of you and what we might be able to learn from those things. And it's really interesting to see everything that's going on out there in the world and watch it being played out on the world stage. It's really very similar to things that are going on inside each of us as individuals.

Rahasya: Yes. As you say, as far as the worldview, sometimes when I watch the news, which is something I try not to do too much, I see the same conflict in my own life on a personal level. The same conflict I have with a sister or friend has the same energetic patterns as some national conflicts.

Shakti Gawain: Absolutely. All our relationships, especially the deep ones, stir up the deepest issues for us that we need to confront and work with.

So I see some of those issues that each of us is individually dealing with as definitely being played out in a big way in the world. And so in addition to just addressing those things externally with whatever solutions we may need, we also need to look inside and see: "How is this reflecting me and what can I learn from this and how can I shift the way I am living so that it has an effect on everyone?"

Rahasya: Looking at it like that, it's like a gift.

Shakti Gawain: Yes, everything that happens to us can be looked at as a gift. Although it's quite difficult when you're in the middle of a hard struggle with something, it's hard to see it as a gift, but in retrospect, we can almost always look back and say, "Oh, I see why I had to go through that."

Rahasya: Do you have any ideas about why the more active type of meditations, like dance and movement meditations, for instance, seem to get better, or maybe I should say, quicker, results in the West than the more traditional Eastern meditations?

Shakti Gawain: Yes, I do have some ideas about that. I think in the Western world we have gotten overly identified with *doing*, and we've kind of forgotten about the art of *being*. And we don't see value in it; we think that if you're not doing something all of the time, being very active and producing something, then you're sort of wasting your time. And there isn't much value given to the necessity for just being quiet. And just resting, and just being, without a focus or a goal. At least a certain amount in our lives—we don't need to do half and half; it's okay if we're doing a lot of doing, we just need some *being* mixed in.

And a lot of us don't have that and we've never really learned how to do that. In many other cultures, and certainly in the Eastern world, there's great value put on being, contemplating, and even withdrawing from the world at certain times or for certain periods of time. But we don't really have that in our culture, so it's difficult for many Westerners to learn how to sit down or lie down and just be quiet without going to sleep. We're just not trained to do it. So I have found also, as obviously you have, that many times people can access that *being* stage more fully and more easily if at first you are actually doing something. If you run, or dance, or do something kind of vigorous, and let the energy release, then sometimes it's easier to sit or lie down and then feel at ease and rest and be quiet and move inside.

But even the very actions themselves can be a meditation. For me, walking is certainly a meditation. First my mind is busy and I'm thinking about all these things and after a while that starts to fall away and I start to become very present in the moment. That's when we know we're connecting with being energy. Even driving a car, I find, driving long distances, I will start to shift into almost an altered state or a quiet being state. And then, oftentimes creative ideas will come or inspirations will come or I'll be able to get a certain kind of breath.

It's been my experience that when one really gets serious about taking a look at one's beliefs, meditation and the practice of simply being present were an indispensable part of the process. For one, it's important to remember while we are looking at the "others" that in fact, they are "us." Many organizations out there are realizing the importance of meditation and being present in regard to our brain's health. One I am fond of is www.thehawnfoundation.org, which is working with and showing that young children today aren't learning the way they should and by being present and showing gratitude they show an increase in neural activity in the frontal lobes. Of course the foundation doesn't call this meditation; it calls it mindfulness activity, probably to escape the censorship of religious organizations.

CHAPTER 29

The Masters of Truth and the Teachers of Beliefs

Truth vs. Beliefs

One of the most important things to remember while reading this book is to not "believe" anything that is written. As Don Miguel Ruiz says in *The Four Agreements* (1997), *"Don't believe anyone."*[70] What he means is simply this: Whatever it is that I know from my own experience as truth can never be directly transmitted to someone else—only alluded to. The best I can do for you is to disillusion you through new information, thereby releasing you from old belief patterns. We all live in this illusion we call life and amazingly enough, we all walk around and miss the most important aspect of our lives, and that's the fact that very few of us know anything at all about our existence.

Most of the time what we run into in life are teachers who believe themselves to be masters and some are outright charlatans. It is rare to have masters appear in our midst, but they do appear and we seldom recognize them and more often than not, we kill them, only to revere them later. The reason is that they threaten the life-dream that we have all come to depend on and recognize as our reality and are supported through our beliefs. Like a person deep in sleep often reacts violently when his or her dreams are disturbed, we as a society do the same. We know to "let sleeping dogs lie" because they tend to bite when aroused from their slumbering sleepy dreams. During World War II the Japanese called the United States "The

70 For more information on Don Miguel Ruiz, visit www.miguelruiz.com.

Sleeping Giant" and didn't want to awake us from our sleep, the sleep we called "The American Dream."

So what do we do? We create immense and powerful organizations to enforce and support our beliefs about the dream-illusion we call our lives and reality. This simply means that we must search deeper, much deeper than what is handed down to us through the mass media by the dream-society in which we live.

Master vs. Teacher

My own search for truth has taken me many places to have many experiences. I traveled throughout Mexico spending time with *correnderos* eating magic mushrooms and later in Brazil with a shaman taking *ayuhaska.*[71] [72] But nothing really compared with the time I spent in India. India in general holds an element of spirituality that has been lost in most Western teachings but India is also missing something that Western teachings embrace. No other person, teacher, or master embraces both East and West as clearly for me as a man named Osho did.[73] In the West you probably know him as

71 *Correnderos* are spiritual guides, much like shamans, who guide and protect you while you are under the influence of an hallucinogen.

72 *Ayuhaska* is made from a root from a plant found in the Amazon jungle and has hallucinogenic properties.

73 On January 19, 1990, four years after his arrest, Osho died, aged 58, with heart failure being the publicly reported cause. Before his death, Osho had expressed his belief that his rapid decline in health was caused by some form of poison administered to him by U.S. authorities during the 12 days he was held without bail in various U.S. jails. In a public discourse on November 6, 1987, he said that a number of doctors who were consulted had variously suspected thallium, radioactive exposure, and other poisons to account for his failing health. Today, Osho's books are more popular than ever before, with translations published in 55 different languages. At the end of the 1980s, the majority of people in South Asia wanted nothing to do with Osho's commune, but since Osho's death, there has been a change in public opinion. In 1991, an influential Indian newspaper, *The Times of India,* counted Osho, among figures such as Gautama Buddha and Mahatma Gandhi, among the 10 people who had most changed India's destiny—in Osho's case, by "liberating the minds of future generations from the shackles of religiosity and conformism." Since

Bhagwan Shree Rajneesh. Probably all you know of him is the propaganda pumped out by the media machine here in the United States because he was a direct threat to the establishment of religion and governments through his teachings.

I want to include what he has to say about beliefs here because his words capture the essence of what this book is about. If I were to attempt to put a label on Osho, I would have to say that he was a "demolition man" sent here to upset and tear down our illusions by demolishing, through pure reason, our beliefs handed down over the centuries by religion and governments. His meditations were also unique; they were a blending of Eastern contemplative and Western bioenergetics and somatic technologies to release traumatic energies held in the body (this is very oversimplified). What follows are some of Osho's thoughts on the differences between teachers and masters and beliefs and truth. Osho, much like Eckhart Tolle and others, was quick to point out that "truth" can only be pointed to but never defined or pinned down.

> The Master is basically a rebel. He is not in the service of the past, he is not an agent of all you can think of as 'the establishment'—religious, political, social, economic—his whole effort is to help you to discover your own individuality. It has nothing to do with tradition or convention. You have to go within, not backwards. He is not in any way interested in forcing you into a certain pattern; he makes you free.
>
> So what I am doing here is not teaching; that is a misunderstanding on your part. But it happens because you have lived with teachers, all kinds of teachers. It is rare to come across a master, because the society does not allow the Master to happen. The society is very afraid of the Master, otherwise why did society poison Socrates? For what? He is the Master *par excellence,* and has never been surpassed by any other. His crime was that he was a Master, and society wanted him to be a teacher. He was helping people to discover the truth. And society is not interested in discovering the truth, it is interested in covering it with more and more lies, because it lives through lies—it calls them beliefs.

---

then, his teachings have progressively become part of the cultural mainstream of India and Nepal. Osho is one of only two authors whose entire works have been placed in the Library of India's National Parliament in New Delhi (the other is Mahatma Gandhi).

All beliefs are lies; howsoever beautifully they are presented, they are lies. Truth cannot be given from one person to another; only lies can be transferred, they are transferable. Truth is untransferable.

The Master cannot hand over the truth to you, he can only create devices so that *you* can discover your own truth. The truth is always your own authenticity, your own being. Who can give that to you? The teacher pretends to give you truth. But what he gives is just a decorated lie—so although it may be very ancient, repeated for thousands of years, it appears to be the truth.

Adolf Hitler, in his autobiography, *Mein Kampf,* says, *"The only difference I know between a lie and a truth is that a truth is nothing but a lie often repeated."* So you become hypnotized by it . . . and you can see it happening everywhere.

People are worshipping stones . . . people with eyes, people with intelligence, worshipping stones! They have been hypnotized from their very beginning. People are believing in all kinds of stupidities, all kinds of superstitions, but they are not aware of it. They are almost in a drunken state; they are living in hypnosis. That is the secret of all hypnosis: repeat a thing again and again and again.

From *Zen, Zest, Zip, Zap, and Zing* (1981)
by Bhagwan Shree Rajneesh (Osho)

If you really want to know who I am, you have to be as absolutely empty as I am. Then two mirrors will be facing each other, and only emptiness will be mirrored. Infinite emptiness will be mirrored: two mirrors facing each other. But if you have some idea, then you will see your own idea in me.
Osho (1931-1990)

Why is it that we crucify and kill the masters who open our minds to our own inner truths and idolize and give our fortunes and blessings to the teachers, preachers, rabbis, and priests who spread, for the most part, lies, deception, and divisive beliefs that have and still are propagating wars throughout history? Why? Because we are asleep in our dream and do not wish to wake up. Why don't we want to wake up? Because the dreamer, all that you dream yourself to be, will not exist in the awakened being that you really are.

That's why it's so difficult for many who will read this book, and others like it, to understand what is being said. To truly wake up requires much more than most people are willing to pay, which from its (the ego's) point of view is everything. Of course the paradox is that by giving up everything you are not giving up anything because it is all an illusion to begin with. What are you giving up in a dream when you awake? You are giving up only the illusion of the dreamer to find yourself in the awareness of what you really are and to again be captured by another level of dreaming. In the Gospels, Jesus says more than15 times that you must awake in this life in the spirit to enter the Kingdom of Heaven. The key here is "in this life" and "in the spirit." It's also important to realize what is meant by the Kingdom of Heaven "within." Of course to even attempt to understand the illusion of past and future goes beyond the scope of what I want to bring out in this book, but it's important to at least point this out because if you can truly live in the "now" you will find that there are no thoughts, therefore no beliefs, no hope, no faith, just pure consciousness, which is where we are headed.

Maturity happens when you start living without hope. Hope is childish. You become mature when you don't project hope into the future. In fact, you are mature when you don't have any future; you just live in the moment—because that is the only reality there is. In the past, religion used to talk about the hereafter. Those were the childish, immature days of religion.

Osho (1931-1990)

CHAPTER 30

Dhara Lemos

Meditation—The Way In and also the Way Out of This Mess

"For me, the practice of meditation provides an experience of what I call the Tao Mind—the aspect of my being that is open, unconditioned, spacious, and intuitive—able to experience life directly. My conditioned mind, that part of my consciousness formed by my memories and experiences interpreted through the filters of

authorities, teachers, family, friends, and other social interactions, is unable to have this direct experience. There is nothing wrong with this conditioned mind. It is just quite limited. Belief is one of the primary limitations of a conditioned mind.

In meditation, we are able to gradually, patiently, compassionately recognize and then lay down all of the conditioned thoughts, opinions, and judgments that seem so real, but actually keep us away from the joy of Life Itself. Beliefs are hard to notice. They are "way down deep" where things are "just the way it is." Meditation is an essential practice if we are ever to see how these "just the way it is" beliefs control and limit our experience of life."

Bill Martin, *The Parent's Tao Te Ching*
(as seen on Oprah's Book Club)

I remember walking through the park one day when a man sitting on the bench I walked by was talking to himself. The conversation was structured around the defense of his ego somehow with statements such as: "They don't know what I know, otherwise they would never . . ." and on and on it went. As I walked by I was thinking, "I'm glad I don't do that." But then I thought, "But I am doing that, the only difference being the fact that I was not verbalizing it." We all have this seemingly endless careening of thoughts streaming through our minds, endlessly attached one to another through association, and have come to "think" of this process as normal because everyone does it. But in reality this is a form of insanity from which we are all separated only by degree.

So what's the remedy? As fate would have it, the ancient meaning for meditation is "remedy." I have to admit that the energy I have spent on practicing meditation has been modest and it seems as though I am always in a state of trying to find the time to meditate. But I know of many people, including myself, who have had profound conscious and what could be classified as metaphysical and mystical experiences while in a meditative state. All too often these states are ignored, discounted as irrelevant, or thought of as the product of a deluded mind.

But from within this inner space of "no mind" or "no thought" while in isolation have come inspirations and ideas that have changed the world. And why is it that being in isolation in a Western society is considered mental punishment and torment? Most people who are in prison with rapists and murderers would rather be there than isolated in solitary confinement. But for centuries, living in isolation has been the source of untold spiritual experiences without which we would most definitely be living in a very different world. It's my guess that the reason we consider this unbearable is that we would be isolated from distractions and be alone with our own insanity that we call our "normal" minds.

Many meditators know what happens when the thoughts die down and "no mind" appears. This is when you come face to face with something else in yourself that most people never know. Of course we could discount this state of mind to be the musings of delusional minds and out-and-out fraud, but that would mean that for millennia thousands of people of high standards and intellect have deliberately taken part in this fraud while at the same time making social contributions for which we owe a debt of gratitude that can never be repaid. No, we need to remove the religious stigma that has been attached to meditation and start looking at it in a more scientific way. What has happened is that atheist groups have all but discounted meditation and the experiences associated with it because of the religious associations it has. Unfortunately, this attitude toward meditation oftentimes makes atheist groups appear just as ignorant as some of the most extreme religious groups.

When I say "no mind," what do I mean? A lot of people "think" that the term "no mind" refers to a mind that simply has no thoughts. This is not the case. It refers to another state of consciousness entirely. It's not part of the mind at all; it's beyond mind. But we are doers and we expect that a state of mind is something to be accomplished by doing, but it's doing that keeps us in the mind. Now what? You could never answer this question in a book but there are many meditation techniques that will help you drop into this "no mind" state.

If you are ready for a teacher, he or she will amazingly come into your life with a little effort on your part; first set your "intention" and then pay "attention." You might want to look into author/teachers such as Adyashanti, Gangaji, and a number of others who practice a process called Satsang, which involves sitting in discussion with others and being guided by someone who is very present (this is an oversimplification). Other techniques involve mantras or simply sitting quietly, such as Vipassana.

However, in our Western culture you may find it difficult, if not impossible, to sit down and try to force your mind to be quiet, even with a technique. But something called active meditations, which I mentioned earlier, have come out of the East and are a blend of Eastern meditation and Western somatic and bioenergetic sciences. This practice was created by masters who knew the problems facing anyone in a Western society who tries to sit still and quiet the mind.

This is the technique I learned while in India and it has helped me in ways that have changed the quality of my life forever. I was fortunate to have met and married someone who has more than 25 years' experience with this technique. Dhara Lemos is from Brazil and has traveled to Germany, England, India, and the United States practicing and teaching active meditations. She now lives with me in northern California where we publish our magazine, the Lotus Guide. So let's hear what she has to say about meditation.

Rahasya: I think when it comes to subjects like meditation it's important to have personal experience along with any research data that is available. It's been shown through fMRI research that meditation in fact does have a positive effect on the brain. This brings the subject of meditation into the realm of topics covered in this book. What I would like to know is a little about your own personal experience with meditation.

Dhara Lemos: My first experience with meditation and how it affects the brain was with a technique I learned from my spiritual teacher, Osho, called dynamic meditation. It is a very intense and active meditation that requires a lot of physical stamina and really challenges the emotions. There is a tendency in most people to do a little bit of the dynamic meditation and stop. But I realized I needed to do this every day for a long time. The first day was easy because it was new. The second day my brain started telling me stories—"I am tired, there is a pain in my back, and so on." But I kept meditating anyway. By the third day, it was almost impossible. My head was hurting and I couldn't sleep through the night. It took lots of energy to go on, but I went on anyway. By the fourth morning something amazing happened. When I woke up, something had shifted in my brain and I was free from the excuses, the fatigue, and the whole "story" that I had been telling myself to avoid the meditation. This particular meditation technique had a profound impact on my brain by changing brain chemistry. This first experience with dynamic meditation was a one-year commitment and meditation became my best friend.

Rahasya: I noticed that you have an outstanding testimony by Dr. James Hardt of the Biocybernaut Institute.[74] He said that your alpha scores were up there with some of the Zen masters he had tested. It seems that the common denominator here is meditation. What are your feelings on your high alpha results in your neurofeedback tests?

Dhara Lemos: It was a great experience to have neural feedback of the results that meditation had on my brain over the years. On the first day of testing I started to do my normal meditation. The audio feedback was loud and high pitched and was an important validation for me. Part of the training is to focus on an issue in your life that you need to work on. So I started with old beliefs about life situations. Then I began to feel forgiveness and compassion for self and others involved in those situations while in high alpha. As I processed the beliefs and my responses to them, I could monitor my brain activity through the sound as well as a numerical score. I still use high alpha in my healing sessions to enable the person I am working with to experience the high alpha state as well. The alpha state is a meditation state where healing can take place.

Rahasya: When I hear scientists like John Hagelin talk about the experiments they have done to reduce crime rates in major cities such as Washington, DC, by getting a designated number of meditators to meditate it made me realize the social significance meditation could have on society. We know now that meditation has a quantifiable effect on the brain in the person meditating, but it also seems to have an effect on the collective consciousness of those in close proximity. How do you think this works?

Dhara Lemos: It's been my experience that when I meditate in a group, it becomes much more powerful than when I am alone. The

---

[74] The process itself is a seven-day training that includes two hours a day inside a closed booth. The brain is hooked with electrodes to a computer so that the brain waves can be monitored. Sound is used as a feedback indicator of the alpha brain state. Dr. James V. Hardt is the author of *The Art of Smart Thinking*, 2007. For more information visit his website at www.biocybernaut.com.

shared meditation increases my awareness exponentially. I had a strong experience in the ashram with a large group of people doing meditation. Upon entering, the air was noticeably lighter; some refer to this as a Buddha field. This makes it much easier to enter the meditation state. Even on a smaller scale, like in my home, I can see that meditation has a positive influence on those around me. So, yes, I believe that this is true . . . if a group or a community is doing a shared meditation, it can totally affect the consciousness of those in the group because we are all interconnected through this conscious field.

Rahasya: How does active meditation work?

Dhara Lemos: When we meditate, we become aware of our connectedness with something greater than ourselves. Dance and music were some of the first meditation forms used by earlier civilizations as a spiritual practice; even now there are many indigenous peoples who still use dance and music as a form of meditation to tap into higher consciousness. From Whirling Dervish to Native American Ceremony, from the Dance of David to the Dance of Shiva, spirit unites motion with form to bring us an experiential fusion with the divine. More recently, in the somatic and bioenergetic sciences we are seeing that we hold trauma in our bodies as suppressed energy. Of course as a child we knew what to do when we were frustrated by life; we jumped up and down and threw a temper tantrum. But the power figures in our lives, our parents and teachers, tell us to go to our rooms and be quiet, sit still, calm down . . . and we do, but at a cost. All that energy gets embodied within us and eventually becomes dis-ease. For instance, right now, sit back and become aware of your body; you will feel that somewhere there is tension that you weren't aware of. This tension eventually becomes chronic if we don't shake it off somehow and let it go. You don't need years of therapy; you just need to release the energy that is blocked. This is how Rolfing, acupuncture, Reiki, and a long list of energy therapies work, by releasing and balancing our energies.

Rahasya: How does connecting with higher frequencies help our lives?

Dhara Lemos: Higher frequencies are at the origin of everything in the universe. They are the stuff of which life is made. This is not just some

New Age jargon; it is hard science. The only difference in anything that exists lies in its signature frequency. This becomes particularly important in the area of health because this is also true for viruses and bacteria. If our frequency is low, we are susceptible for interaction or what scientists would call "entrainment" with those frequencies that can harm us. By allowing the frequencies to move through us, we become like antenna to receive more of the "good stuff" available to us as we cocreate the relationships and circumstances of our lives.

Rahasya: How would meditation help in looking at our beliefs?

Dhara Lemos: It's the active element of meditation that brings about a change in chemistry that brings about a change in mental processing. Beliefs are supported in part by their chemical counterparts and when you change the chemistry you might say you reformat your hard drive.

Through the years I have observed the change in many friends who have practiced these meditations. They require a commitment that many people simply are not willing to make. For one, when you start practicing any meditation, especially these, you will stir up everything that you have worked hard to let settle, either through drugs (both legal and illegal) and escaping through entertainment and other means.

If you are interested in knowing more you can contact Dhara through my website, which is listed at the end of this book.

CHAPTER 31

Dhara Lemos, Dr. Joe Dispenza, and Rahasya

So What's the Solution? "Evolve Our Brains"

Change your thoughts and you change your world.
Norman Vincent Peale (1898-1993)

Again, we will never come up with the solution from the same level of awareness that created the problem in the first place. We need to free our minds and "evolve our brains," as Dr. Joe Dispenza would say.[75] This is where we get into giving serious thought to how we limit

[75] *Evolve your Brain: The Science of Changing Your Mind,* 2008.

our capacity to think and reason by having limiting beliefs that are rigid and dogmatic. There is wisdom in analyzing our deep-seated beliefs in a scientific manner, and any believer should embrace this system because some beliefs may be true. And if they end up not being true, which historically speaking, most are not, then what kind of consciousness would want to hold onto them? Imagine living in today's world still believing the world is flat so you can not go too far for fear of falling off the edge. Or that a woman who uses herbs for healing is by definition a witch and should be burned at the stake.

We obviously need a new way of perceiving the world and that comes about through a new level of consciousness. We know now that the old belief that our life is predetermined by our genes is not true.[76] Cells can live for quite a while without any genes at all; they react to their environment, communicate with other cells, and continue to live. We also know our genes are affected by the environment through interaction with the cell membrane. Nothing seems to be predetermined any more, which means we can change our bodies and our minds through thought. For many years we were led to believe that "nature" determined our lives, which basically put us in the role of being a victim in this world. Now we are starting to see through scientific inquiry and research that it is "nurture" that plays the dominate role in our behavior, health, thinking, and consciousness.

In my recent conversation with Dr. Joe Dispenza, well-known for his part in *What the Bleep Do We Know?*(2004) and author of *Evolve Your Brain* (2007), we touched on issues that were once taboo to even think about and that are now entering mainstream thought.

A Conversation with Dr. Joe Dispenza

Rahasya: Why is it that most people cannot break out of and change their perceptions or beliefs?

Dr. Joe Dispenza: It really doesn't matter which one we are talking about, perceptions and beliefs fall into the same category here. You can even include attitudes and habitual thought patterns. What the brain likes to do is to simply replace old ideas or beliefs with new ones. Basically there is an exchange of self-limiting thoughts or beliefs for new and, hopefully,

76 *Biology of Belief,* by Dr. Bruce Lipton, 2008.

improved ones. It's actually far better to interrupt negative thoughts and beliefs by creating a new level of awareness which simply allows room for more positive thoughts and beliefs that may serve you better. This actually rewires the brain in such a way that it changes behavior.

Rahasya: When you say, "New thinking and new beliefs can literally rewire one's brain to change behavior and habit-forming patterns," it's easy to see that this could have profound implications, both individually and collectively, but how do we, not just change, but let go of something that we have identified with so heavily, such as our beliefs, whether they're social, political, or religious beliefs?

Dr. Joe Dispenza: "First of all, let me describe new thinking. When I talk about new thinking, I'm talking about "new ways of thinking." For the most part, we inherit many different patterns of synaptic connections that formulate a lot of our unconscious thinking. These patterns of *neurological wiring* basically comprise and are developed by what our parents have emotionally embraced. Those unconscious thought patterns are really the result of what our parents have repeatedly thought, consistently done, and finally experienced.

When we are born and when we develop into adulthood, those patterns seem to be pretty much our standard ways of thinking that are set in place for us to overcome. So, the routine automatic programs that we run our life by are pretty much genetically predetermined. Then we're conditioned from our environment to act and behave in certain ways. The combination of genetics and environment contributes to the belief systems, behaviors, and perceptions that we live by on a daily basis. Then we begin to gain information and knowledge, and we begin to ask questions in our present understanding that challenge the current belief to see what works and what doesn't work and then begin to investigate new ideas and new paradigms. And to learn that information and then remember it causes new series of patterns and connections to form in the brain, which creates literally new levels of mind. And to be able to repeat those levels of mind at will, to be able to remember and then apply what we've learned, and apply what information we've embraced, begins to set up new experiences.

It's not enough just learning and creating new ways of thinking. It's actually taking that information and applying it; the application sets up a new sensory experience, which then changes our belief system. It alters the way we're perceiving, responding, reacting, and ultimately believing. And that takes practice and that takes will and that brings a certain amount of discomfort in most people's lives."

Rahasya: And since we spend most of our energy and time either avoiding pain or discomfort or pursuing pleasure, this would explain one of the reasons our beliefs continue generation after generation even in the face of facts to the contrary.

Dr. Joe Dispenza: Yes. And you know, looking at evolution, any species that makes an attempt to evolve has to be confronted with some environmental circumstances. And it could be climate, it could be social pecking orders, it could be predators, but whatever it is, there is some discomfort, chemically, that causes the creature to actually develop a new cognitive process, to formulate a new way of being. And that cognitive process, essentially, is encoded in their tissue and that's what evolution is about. The discomfort is what pushes the envelope of the familiar habits and beliefs and forces them to come up with new ways of being, new ways of thinking, new ways of reacting, faster legs, camouflage, migratory patterns, whatever it is that then allows them to actually become greater than their environment. My book is an attempt to help people to begin to understand that you can create a mind that's greater than our environment, that's greater than our conditions, and as human beings our conditions tend to be a little bit more complex. Whether it's social situations with your boss or your family, whether it's disease, whether it's traffic jams, Internet connections, raising teenagers, single parents, whatever it is, those are all equally environmental conditions that we have to drive ourselves to get better at mastering. And that in turn is also encoded in our genetics. It begs the question, "Does the environment control my thinking or does my thinking control the environment?"

Rahasya: When you say we become addicted to familiar patterns because we are not being stimulated by new experiences, could you explain what kind of experiences could break, let's say, old

dysfunctional religious beliefs that no longer serve us but nevertheless have us in their grip?

Dr. Joe Dispenza: When we start talking about religion we're talking about indoctrination—accepting ideas as the final word without looking and asking deeper questions of inquiry into those beliefs. It was the book that was created, that was passed down through generations, that was God's word. It's very difficult for people who want to believe in something to change and go against that doctrine or that structure because that literally puts them in the position of where they could actually be banished from God, or separate from God. So it takes true individualism and courage to ask those big questions. It's through leadership and free will that you can begin to combine common sense with empirical information and create new models outside the boundaries of accepted beliefs. But it's also important to actually apply, to actually try out and validate that piece of information that was passed down or that piece of information that is being investigated.

Rahasya: Well, to be truthful I think we have been trying out that piece of information and to tell you the truth it doesn't seem to work. Here's something that I've been thinking about lately. As I've looked around in my own world, ever since I was a child, I've noticed that people with rigid beliefs have a difficult time. They may even have a high IQ and be well educated. But understanding anything outside the parameters of their particular belief seems to lead them to the point where some of them can commit some pretty outrageous acts. For instance, 14 of the 19 people who flew planes into the towers (if we believe the official version of what happened) were highly educated, from good families, yet their beliefs dictated to them to do something that most of the world would consider preposterous and absurd, not to mention immoral. So what I'm wondering is, do you think this kind of thinking, this kind of believing, this kind of mental state could create neural networks that are not only inconsistent with evolving your brain but that could actually inhibit what has become known as spiritual awakening?

Dr. Joe Dispenza: That's also a very good question and the answer is *definitely yes*. I think that fanaticism has been going on for as long as man has walked on two legs. And I think that intellectual ability or learning has nothing to do with spiritual aptitude, emotional intelligence, or emotional ability. Emotional intelligence fundamentally means having our thoughts control our emotions, not having our emotions control thinking. Most people who make choices accordingly may have a certain degree of intellectual maturity but are very emotionally immature. And I think that fanatics tend to see certain causes and then they rationalize their choices emotionally using their beliefs and then act accordingly. And that in a sense is a very dangerous situation, because they're feeling very righteous in their actions to accomplish something that they think is going to promote the advancement of a certain belief system. And yet it harms many people in the process, and that's been going on in cultures throughout history. With all the progress that's happened in this time period you'd think that people with any degree of awareness would see this. But let's face it, all of us as individuals have organized personal mind-sets that have become our present ways of perceiving and being in this world. And we've been influenced by religion, culture, genetics, upbringing, and education. That's pretty much formulated our personal identity and we use what works but at the same time we keep using what doesn't work also. So this would be a good opportunity to self-reflect and ask, "Where am I with my own personal understanding of myself and my relationship to the world, and what do I need to let go of, to not only be a better person but to be able to influence the world around me in a positive way?"

Rahasya: For me, what has helped is letting go of the need to believe in beliefs. This has become more than an intellectual curiosity for me; it has become a spiritual passion and quest to ask these politically incorrect questions. It seems when we look out at how society is evolving, we're actually confronted with a choice. And the choice seems to be to live by beliefs that have been handed down to us from generation to generation that no longer make sense to anyone who is being intellectually honest with themselves, or to come to grips with our mystery and learn how to live with it and in it. And the only way, I think, we're going to learn how to live within the mystery of life is to somehow evolve our brains and our minds. Then when we reach the limits of our knowledge and understanding we won't fill in the rest with beliefs of what we wish or hope to be true. Then we could be content with the slow unfolding of knowledge through an honest intellectual search.

Dr. Joe Dispenza: Yes, and it's important to remember that the natural knee-jerk response to those conditions is for us to begin to judge and begin to react. And I think, for me personally, that it's not going to create any solutions on an individual level. I think that when we begin to become self-aware and self-reflective and see that that pattern is within each and every one of us, and then begin, if you will, a spiritual awakening and begin to say, "What do I need to change about myself?" Then begin to eliminate anything that ultimately resembles these states of mind or patterns in myself. And we can't judge the rest of the world for being aggressive in their religious beliefs and at the same time cut people off on the freeway and expect to think that it's not the same mentality. And so, taking it in and rising above what we're presented with in life is the true art, the true idea behind awakening to higher consciousness.

There's a second point about this, and that is that the one thing we know is that free will is a God-given right of every human being. And there are going to be people who for long periods of time are going to continue to think the world is flat no matter how much empirical evidence there is to the contrary. And changing beliefs on a global level is, as history proves, not an easy process. So, the first people who think outside the box, and are interested in new ways of being, will be considered the mavericks, and they will be pointed at as fanatics by the very people we call fanatics, and then that test of uncompromising individualism is what will change the course of history. Because history has shown that whether it's Martin Luther King or William Wallace, any great charismatic leader confronts the conventional beliefs of the time and is met with resistance from those in power. They're willing to put their life on the line for a principle or an idea that is bigger than they are. The people on the planet who are embracing new ideas and understanding will evolve their brain and this is what will define and alter the course of history on a global level. The real experience of living it and being uncompromising as individuals will create a new world, which I think is much needed in our present time.

Rahasya: Yes. One of the things I learned from reading* A People's History of the United States *by Howard Zen is the fact that most great movements that have ever happened on this planet have never happened from the top down. They've always happened from one person having one idea and starting a grassroots movement, so

it happens from the bottom up. And it seems as if we historically have tried to give all our power over to governments and large organizations to accomplish these great changes for us. But when it comes down to it, it's up to us as individuals to do this. It's probably to our advantage that the very same organizations that would try to stop us are the very same organizations that seem to be dismantling themselves. The reason I say dismantling themselves is because it's their actions and transparency that are doing most of the dismantling, not our laws and regulations.

Dr. Joe Dispenza: And I think that the disease, or the plague, in this country right now, and in the world, is really the convenience of complacency. It's very easy to be entertained. It's very easy to hit the power button on the remote control, or surf the Internet, and see something disgustingly horrific that threatens millions or hundreds of thousands of people and then change the channel and in a matter of moments become absorbed in something else and completely forget about what was so disturbing moments before. And the quick changes in attention because of our media structure and technology are enough to subdue and hypnotize people from really making strong choices that are going to be uncomfortable.

Dr. Joe Dispenza is one of many researchers in our world today who encourage us to ask the tough questions. I recommend reading his book so you can see for yourself the vast empirical evidence that scientists are compiling on the subject of evolving our brains through breaking down old patterns and belief systems.[77]

---

[77] Dr. Joe Dispenza studied biochemistry at Rutgers University in New Brunswick, New Jersey. He received his Doctor of Chiropractic Degree at LifeUniversity in Atlanta, Georgia, graduating magna cum laude. Dr. Dispenza's postgraduate training and continuing education have been in neurology, neurophysiology, and brain function. Dr. Joe was a student and a teacher in Ramtha's School of Enlightenment, and he was one of the scientists interviewed in *What the BLEEP Do We Know!?* (2004). His DVD series, *Your Immortal Brain* (2005), looks at the ways in which the human brain can be used to create reality through the mastery of thought. His new book is *Evolve Your Brain* (2007). For more information, visit www.drjoedispenza.com.

Part III:
Rewriting Our Human History

CHAPTER 32

In the Beginning, There Was Light

In the beginning light filled the void; that may be the first thing that science and religion have been able to agree on. I would like to take a little time to explore some of the things we know to be true about our place in the universe. Of course, this isn't entirely scientific because there are many unknowns but we can make some fairly safe assumptions based on what we do know and observe today. At least we have some historical evidence and hard science to lean on to make our extrapolations and form ideas. So let's go on a little adventure and see where we end up.

Some say it was 12 billion to 14 billion years ago, others say it's closer to 18 billion, but that really doesn't matter for what we want to explore. It appears that in the beginning, based on the Big Bang Theory or Standard Model of the Universe, there was a release of energy far beyond what we can imagine where matter, time, and space itself manifested and expanded.[78]

---

[78] In physical cosmology, the Big Bang is the scientific theory that the universe emerged from a tremendously dense and hot state about 13.7 billion years ago. The theory is based on the observations indicating the expansion of space in accord with the Robertson-Walker model of general relativity, as indicated by the Hubble red shift of distant galaxies taken together with the cosmological principle. (From *Delusion Addiction: God-Tyranny-Disaster* (2007), by Bridger Daquan, pg. 32.) Extrapolated into the past, these observations show that the universe has expanded from a state in which all the matter and energy in the universe was at an immense temperature and density. Physicists do not widely agree on what happened before this, although general relativity predicts a gravitational singularity. It might be interesting to mention that in the past few years astronomers have observed more than 2,000 explosions at the edge of our known universe that are equal or close to the type of explosion that's

[79] In those first few moments all the energy of the known universe was released and was so condensed and hot that nothing was formed; everything was in a quantum state of energy and not even what we know of as subatomic particles existed in those early moments. Slowly, everything cooled down and hydrogen atoms started to form, along with a small percentage of helium (approximately 5 percent). Discoveries in astronomy and physics have shown beyond a reasonable doubt that our universe seems to have a beginning; even the word "beginning" loses its meaning at this point because before that moment there was nothing, no time, no space, no-thing. During and after that moment there was something; that "something" is what we now call "everything." The Big Bang Theory is an effort to explain what happened during and after that moment.

According to the Standard Model, our universe sprang into existence as a "singularity," which defies our current understanding of physics. Singularities are thought to exist at the core of "black holes." Black holes are areas of intense gravitational pressure. The pressure is thought to be so intense that finite matter is actually squished into infinite density and the only thing left is the gravity from the matter that once was there (a mathematical concept which truly boggles the mind).

Whatever caused that event, the Big Bang, in the first place has and may always remain a mystery, but what is not so much a mystery has been what has happened since. From that moment it seems as though there has been an underlying consciousness coming in and out of form through more and more complex patterns of being, forming cooperative units, with those units then coming together to form larger and even more complex systems by connecting and forming even larger and even more complex systems. Within the primordial soup of subatomic particles came our first elementary atoms, hydrogen and some helium. Then through gravitational influences the first stars were born and with each exploding star, or supernova, came the first heavier atoms, which formed simple molecules. And from those simple molecules came the first one-celled organisms, which slowly evolved

---

predicted in the Big Bang Theory. It would be difficult to predict what this may mean but one thing is for sure; the universe is far more complex than we have ever been able to imagine.

[79] This also agrees with much of the Hindu religion, in which its practictioners referr to an oscillating universe where there was no first time and will be no last time.

into even more complex creatures until finally that very same consciousness became self-aware upon this tiny little planet of ours in the form of human beings, one of whom is reading this book right now. This points strongly to there being a pattern of some kind of force that has the intention of becoming more and more conscious through pulling smaller parts of its manifestation together into larger and more complex organisms through attraction and connection.[80] It takes so many atoms to make a molecule and so many molecules to make a cell and so many cells to make an organism, such as us, for example.

It's been only in the last several thousand years that our brains have reached the necessary complexity for self-awareness as individual human beings. This is the point where the discussion gets interesting, as if it's not interesting enough already. If this is indeed the intention of universal consciousness, to become more conscious through coming together into higher and more complex states of being and uniting into a larger and more complex organism, then there is no reason to think that we are the top of the pyramid of this conscious evolution. If the process continues as it has for billions of years, that means the next step in our conscious evolution would be to do what we have always done—come together to form an even higher and even more complex organism. Of course the alternative to this theory is that consciousness has finally reached its peak manifestation in the form of human beings. I doubt we could even seriously consider that we are the highest forms that creation is capable of producing. That would be the epitome of human arrogance to even think so. The only difference is that now we have evolved to the point where we are conscious participants in this process of conscious evolution.

With our participation what was once the evolution of consciousness has become conscious evolution.
Rahasya Poe

[80] I see this "attraction," even at the very beginning of the universe, as what we could define as "love." Our insatiable need to attract and connect with others could be the primary reason and process that the universe used to come in to existence in the first place. But this is one of my beliefs, so think it out for yourself and come to your own conclusion. The process of coming to a conclusion is what the journey is all about, not just accepting other beliefs blindly.

So, if this pattern of bringing smaller units together to form larger and more complex units is true, it follows that we will come together as individual human conscious forms (planetary cells) to unite consciously as a planetary being in which we, as humans, connect with each other and transform into a higher, more complex being, a being that would be as difficult for us to imagine as it would be for a cell in our bodies to imagine the complex human organism that it is part of. Indeed, this seems to be what has been happening now for several years in our outer world. We have experienced a constant connecting through technology, earlier in the form of postal services allowing people to transmit or share their thoughts over space and distance, and then telegraphs and telephones. Of course it's easy to see that now we are all connected on an outer dimension all over the world through the Internet. This connecting can also be observed through the coming together of the United States and more recently the EEU in Europe. As of this writing, there are plans to unite the United States with Canada and Mexico. When you think of how the printing press changed the world by letting people into other people's minds from different times and places and consider what the Internet is doing it's easy to see that we are fast approaching a time when we will enter into a new dimension of thought and consciousness. And this is only the outer manifestation of what's happening on the inner planes of our spiritual existence.

So what does all this mean to us in our individual little world of thought? It means everything. We are at a crossroads where our individual ego personalities are fighting with every trick in the book to hold on to their individual identities. The situation is that a lot of people are waking up all over the planet and putting a lot of psychic stress on ego personalities—individual, collective, and organizational. This in turn is causing everything to heat up through conflict. One of the best tactics an ego has is to form groups based around some type of ideology or belief that it is special, thereby making the group special. This, of course, causes conflict because the only way you can be special is for someone else to not be special. It's been easy to see that being chosen by God as special is not an easy role to play out in the world without generating a great deal of conflict and suffering.

Conflict is the best way to crystallize your ego personality because it's almost impossible to be in the here and now while in conflict. To be in the here and now requires "oneness" and "stillness." Ego cannot "be" in that state; it needs time, both future and past, to exist. It exists through thoughts that breed negative emotions such as anger, aggression, fear, remorse, guilt, suffering, feelings of lacking what you need to survive, anything to pull you out of "now."[81] It's not an accident that religious and nationalistic beliefs are divisive; they are the constructs of an ego personality for the explicit purpose of coming together in ever larger groups by rejecting and, if need be, eliminating those groups who believe differently.

It might also be worth noting that our negative emotions suppress our neural connections through chemistry. Once this happens you find yourself in a place where it's difficult to understand new information. A mountain of historical evidence backs this up. In the 50 trillion to 80 trillion cells of our bodies there are more than 100,000 chemical reactions every second, and through the new science of psychoneuroimmunology and epigenetics we know there is a deep connection with mind and body and the environment. Eventually, you always end up thinking the way you feel and feeling the way you think. I remember reading this in an old book: "As a man thinketh in his heart, so is he."[82]

It's easy to see that this is why individuals come together in groups through the sharing of strong emotional ties and beliefs; they share a chemistry that needs support. The ironic part is that the process not only gets support through others who agree, it also gets support from others who disagree, so as a group that has a shared illogical irrational belief we always seek out enemies to ensure our ego existence.

Noosphere or Biosphere (Planetary Consciousness)

So here we are on our continuing conscious evolution through time and space. The idea of a collective conscious has been around for quite some time, but it was popularized in the last century by Pierre Teilhard de Chardin, a Jesuit priest and a very controversial theologian who also

---

81 Read *The Power of Now* (2004) by Eckhart Tolle for an in-depth look at this.

82 Proverbs 23:7. This verse is often misquoted and actually reads: "For as he thinketh in his heart, so is he: Eat and drink, saith he to thee; but his heart is not with thee." This concept is also found in most Gnostic writings. This is one of the many statements to be found in ancient writings that lead one to understand there is a secret hidden knowledge that was meant to be passed on from generation to generation. The task for us is to clear away the nonsense.

happened to teach chemistry and physics.[83] He believed there is a global mind, a collective consciousness, or what he called a "noosphere," which is part of our evolutionary path toward conscious awakening as a larger universal being. He saw this as part of our planet's ecosystem and similar to what we might call our biosphere. On this subject, you may want to give some thought to the real meaning of being "part of and one with the body of Christ" and Christ Consciousness.[84] This is one more example of peering deeper into the esoteric meaning of sacred scriptures, but this can be done only when we let go of the traditional and literal meanings into which we have been indoctrinated.

Indeed, there have been numerous experiments in which researchers have observed that even on an atomic level, atoms stop acting individually when they are in a high density of particles (plasma). Experimental physicist David Bohm became interested in collective consciousness after observing this phenomenon at the Lawrence Radiation Laboratory in 1943. He commented that it was as if the electrons were alive.[85] This was one of the first clues for Western society that our universe is deeply connected. Of course this is the hallmark of Eastern spirituality.

---

83 Pierre Teilhard de Chardin was a French Jesuit priest trained as a paleontologist and a philosopher. Teilhard conceived such ideas as the Omega Point and the Noosphere. Teilhard's primary book, *The Phenomenon of Man* (1955), set forth a sweeping account of the unfolding of the cosmos. He abandoned traditional interpretations of creation in the Book of Genesis in favor of a less strict interpretation. This displeased certain officials in the Vatican, who thought that it undermined the doctrine of original sin developed by Saint Augustine. Teilhard's position was opposed by his church superiors, and his work was denied publication during his lifetime by the Roman Holy Office. Pope John XXIII rehabilitated him posthumously, and, since then, his works have been considered an important influence on the contemporary church's stance on evolution. From Wikipedia

84 1 Corinthians 12: 26: If one part suffers, every part suffers with it; if one part is honored, every part rejoices with it. 27: Now you are the body of Christ, and each one of you is a part of it.

85 *Infinite Potential: The Life and Times of David Bohm* (1997). David Bohm was an excellent example of someone using intuition to guide himself and science to validate the journey. He questioned a lot of the scientific orthodox beliefs of our time and was, of course, admonished by his peers for doing so.

We are all linked by a fabric of unseen connections. This fabric is constantly changing and evolving. This field is directly structured and influenced by our behavior and by our understanding.
David Bohm, quantum physicist (1917-1992)

Later Bohm went on to study consciousness and human communication in particular. One of his conclusions was that a great deal of human misunderstanding arises out of the fact that people simply do not hear others when they are communicating. What he found is that he could improve dialogue by getting those involved to, at least temporarily, *suspend their beliefs* to be more receptive. This became known as Bohemian Dialogue. Participants in his groups would report the feeling of a group consciousness that they could all tap into for creative solutions. The process of suspending your beliefs or putting them aside is also highly recommended while reading this book

So exactly what am I getting at here? What does this have to do with personal beliefs, especially divisive beliefs that declare us as an individual to be unique or chosen somehow as special? We've all observed this tendency toward a "group mind" in nature. We see it with "flocks" of birds, with "schools" of fish, with "swarms" of bees, with "packs" of wolves. They are obviously connected somehow in ways that aren't entirely clear. But what is clear is that they are in a collective mode without which they could not survive. For a moment, imagine a flock of geese flying south, in perfect harmony, one taking the lead when the other gets tired, not for leadership or some personal gratification in claiming its dominance or superiority over the others, but merely to take its place in the scheme of things.

Now imagine if there were, let's say, three different groups of geese that had conflicting and divisive beliefs about themselves and the other geese. Imagine also that each group has the belief that it is special and chosen by the invisible Goose God to dominate the others. Part of the conflict is which goose and which group will lead because whichever goose leads gets special privileges and positions, even within its group. This would obviously break the collective bond each group had with the other two groups and all of a sudden it would no longer be about flying and getting to their destination; it would be all about ego and competition.

Of course the analogy is irritatingly obvious, but if this is true with us, what amazing visions of a new world are we missing by maintaining these outdated and dysfunctional beliefs created by the egoic mind thousands of years ago in an era of abject ignorance of the world around us? We are constantly trying to prove something, not only to ourselves but to others, and in the process we lose sight of the very miracle that we are part of and the unfolding of miracles that we can not even now imagine that lies ahead of us. This is why it is so important to understand the concept of spiritual or conscious evolution and how our beliefs could be inhibiting an important phase of our evolution.

If God did not exist, it would be necessary to invent him.
Voltaire (1694-1778)

Now let's look at this analogy one more time and imagine that there's only one goose instead of groups of geese that believes he or she is somehow special and therefore deserving to have his or her way at the expense of the group. I doubt if the single goose would have much impact on the group and would, in the end, be left behind and expelled by the group for the greater good. So what would this lone egotistical little goose need to do? It would need to get others to believe the way it did and to do that it would need to come up with a group ideology or belief into which the group could be enlisted and identify with.

It is in this way that the collective ego gains power, and the power it has as a whole is much greater than its parts. Of course this would eventually cause a division within and from the larger group, which is the beginning of an ideology, belief system, religion, or nation. It's also important for the group to believe the "belief" of the day and the best credentials for that is "God spoke to me." Of course the Goose God would be speaking from a white cloud, not a burning bush, where unfortunately no one else happened to be at the time, which always seems to be the case.

But fortunately, geese don't seem to have the thought processing necessary to develop these tendencies the way we do as human beings. Once our brains reached the complexity needed for self-awareness and the development of ego, we started breaking away, not only from the group but also from the world in which we live. This is what Alan Watts calls the "skin-encapsulated ego," in which we say, "I am this, and I am not that."

We seem to be on the cusp of a new way of looking at our relationship to the world we find ourselves in. When I talked with Peter Russell about this he had a lot to say because this is in line with his Global Brain Theory or the Gaia Hypothesis.[86] He wrote the following article for my magazine, very eloquently titled "From Science to God: The Journey of a Devout Skeptic."

---

86 Peter Russell wrote *The Global Brain* (2005), which was also made into a video that's available on his website at www.peterrussell.com.

CHAPTER 33

From Science to God: The Journey of a Devout Skeptic

By Peter Russell

That we are conscious beings is the most obvious fact of our existence. Yet as far as Western science is concerned, there is nothing more difficult to explain. Why should the complex processing of information in the brain lead to an inner personal experience? Why doesn't it all go on in the dark, without any awareness?

This paradox—namely, the undeniable existence of human consciousness set against the absence of any satisfactory account for it—suggests that something is seriously amiss with the contemporary scientific worldview.

I believe that rather than assuming consciousness somehow arises from the material world, as most scientists do, we need to consider the alternative worldview put forward by many metaphysical and spiritual traditions in which consciousness is held to be a fundamental component of reality. When we do, everything changes, and everything remains the same.

As far as contemporary science goes, nothing is lost. Mathematics doesn't change, nor do physics, biology, or chemistry. What changes is our understanding of ourselves.

Those who have devoted themselves to a personal exploration of consciousness have repeatedly discovered a deep inner union with the divine, expressed most radically as "I am God." To traditional religion, this sounds blasphemous. How can any lowly human being claim that he or she is God, the almighty, supreme deity, the eternal creator?

Meanwhile to modern science, the notion of God is unnecessary. Science has looked out into deep space to the edges of the universe, back into "deep time" to the beginning of creation, and down into "deep structure" to the fundamental constituents of matter. In each case science finds no evidence for God, nor any need for God—the Universe seems to work perfectly well without any divine assistance.

When mystics refer to God, they do not speak of a separate being existing somewhere in the physical realm; they point toward the realm of personal experience. If we want to find God, we have to look within, into the realm of deep mind—a realm that science has yet to explore.

The worldviews of science and spirit have not always been as far apart as they are today. Five hundred years ago, what science there was existed within the established worldview of the Christian church. Following Copernicus, Descartes, and Newton, Western science broke away from the doctrines of monotheistic religion, establishing its own atheistic worldview, which today is very different from that of traditional religion. But the two can, and I believe eventually will, be reunited. And their meeting point is consciousness. When science views consciousness as fundamental to reality, and when religion understands God as the light of consciousness within us, the two worldviews start to converge.

Today this meeting of science and spirit is critical, not just for a more comprehensive understanding of the cosmos, but also for the future of our species. We desperately need a worldview that validates spiritual inquiry, for it is the spiritual aridity of our current times that lies behind so many of our crises.

When I first watched Peter Russell's video documentary *The Global (1987)*, it made total sense to me and I highly encourage you to get this video and watch it with friends. It will take you through our evolution as consciousness, how we came to arrive at our present moment, and where this may be heading now that we are starting to play a more conscious role in our own evolution, which is the good news and the bad. It's bad only if we hang on to beliefs and worldviews that no longer support this new awareness arising in us as a global or planetary being. We would do well to remember that extinction usually comes about through the inability of a species to adapt. For more information on Peter Russell visit www.peterrussell.com.

CHAPTER 34

Dhara Lemos, Dr. Emoto, and Rahasya

Hidden Messages in Water
An Interview with Dr. Masaru Emoto

For those who may not be familiar with Dr. Masaru Emoto's work, Dr. Emoto captures frozen water crystals in photographs that present a glimpse into water's mysterious response to thoughts, words, and pictures. You may know him from the movie *What the Bleep Do We Know?* (2004) in which it was brought up that if our thoughts have this kind of observable effect on water, what kind of effect do we have on ourselves since we are made mostly of water? The answer is obvious—we have a profound effect on our physiology with our thoughts and emotions. If this were all there was to think about it would be profound, but when you add to this the discoveries in other branches of science we see that we are coming face to face with our own cocreative powers. For instance, it has been shown through research that we have neurons in our hearts that actually communicate with our brains through a neural network.

When I met Dr. Emoto, I was impressed by his simplicity and ability to explain his work to most everyone who listened, including children. I would like to encourage you to visit his website and see more photos and information. Researchers such as Dr. Emoto are important because they are stepping out of the boundaries of what we believe to be true, but they are validating each step and each step takes us deeper into the realms of pure consciousness. Here are some of the highlights of our interview with Dr. Emoto.

Rahasya: I think "validation" is important because it allows our minds to let go of questions and limitations and go on to other levels of awareness. What is it that your work "validates" and how does this help us in our everyday lives?

Dr. Masaru Emoto: A familiar example that I believe my research "validates" is the idea that "the water holds and transmits information." Also, I could say this is validating the proverb "The water is the origin of all things," as quoted by Thales. As you know, water changes its dimension from solid matter to liquid matter and to gaseous matter (and probably to light at last). Thus, the more we properly understand water, the more we can expand our consciousness into other dimensions.

The validation for me personally has been invaluable because to be honest I could not sense the invisible world through my own sensibilities. When people would speak of the inner world of our psychology I would wonder if that was even possible, but through seeing how our thoughts have an observable effect on water I no longer doubt and it's because of this validation.

Rahasya: Do you think research like yours indicates our expanding consciousness into other dimensions?

Dr. Masaru Emoto: Understanding that water exists as a solid, a liquid, a vapor, as a molecule, and at some point as light is to also understand our own nature and possibilities because we are mostly made of water. After I realized that we are water I came to realize that we do not have death; we only change into higher vibrations of being. This is an irreplaceable finding for me personally.

Rahasya: It seems to me that a lot of research has been going on with how our thoughts affect our DNA, which in turn has a direct effect on our health. From what you were saying in your lecture I gather that you believe water, even though it is not self-aware, holds and transmits our thoughts and emotions. Do you think our DNA also works along these same principles?

Dr. Masaru Emoto: Actually, I regard DNA as "little, but strong, water" so yes, I think so.

Rahasya: What would you like to see come out of your work?

Dr. Emoto: If each one of us starts thinking, learning, and understanding more about water and the fact that we are water-based ourselves, then we can start to love, thank, and respect water in our daily lives.

After the interview Dr. Emoto and I talked about belief systems and how they are dividing the world. His feelings are, without a scientific view of the world, that we will be forever caught in the confusion of conflicting belief systems and the only way through this will be through scientific validation and research.

For more information visit www.hado.net.

CHAPTER 35

Dhara Lemos, Amit Goswami, and Rahasya

Reconciling Evolution and Creation
A Talk with Amit Goswami

Amit Goswami is an extraordinary quantum physicist and visionary thinker. I first became aware of him through watching the documentary movie, *What the Bleep Do We Know?* (2004) I think that some points that come up in our conversation would be more complete if we had taken the Sumerian Tablet/Anunnaki information into account but I avoided bringing it up because I wanted to hear his point of view in its pure state. I am happy with this choice because what he has to say is a beautiful way to look at both evolution and creation theories and paints a very clear picture of how we could start reconciling these two points of view without bringing in the Sumerian information; the Sumerian information is simply too much to ask of some people when they are still deeply embedded into a belief system. What's interesting is that when you understand what he is saying and stand back and look at it, you can see that even though he paints a clear picture, a piece is still missing. We will be covering the Sumerian Tablets and the Anunnaki information in more detail later and you will see how this information possibly ends up being the key to understanding and explaining the missing link in our evolution and to a more complete understanding of creation.

Rahasya: In your book* The Visionary Window *you say consciousness creates life and guides its evolution and in the process it resolves the intelligent design/neo-Darwinism controversy. This would be a controversy well worth resolving; could you explain this a little?

Amit Goswami: Yes. Darwinian evolution is slow and gradual, step by step. Such an evolution can explain microevolution, but not macroevolution, when a large number of components change, as in the development of a new organ; for example, how did the eye evolve? The idea behind Darwinism is that organisms adapt, and that nature selects only those genetic changes which are the mutations that serve a good purpose for adaptation. So taken this way, the eye cannot develop gradually because 1 thousandth or 1 millionth of an eye would be of no value for survival. So generally this question rules out Darwinism as an adequate theory for macroevolution.

Intelligent design is a good idea in the sense that it makes sense of the idea of purposiveness in biology, which biologists have denied for a very long time. But that denial is based more on the inability of modern biology to explain purposiveness than to disprove it. Anybody who looks at a living organ sees that it's built for the purpose of performing a function. If we see a kidney, it is impossible not to see the purposiveness of removing the garbage from the body to keep the body functioning properly. So in this light intelligent design looks fairly attractive; however, it is not scientific in the sense that it does not fit into a scientifically verifiable part of a theory. It just concludes that God does it and God does it without evolution, but there's an enormous body of scientific evidence in favor of evolution.

What I do with the help of the new science within consciousness, and that quantum physics is allowing us to develop, is to show that the fossil record has a distinct record of very fast epochs where evolution is very quick. I found quantum leaps in evolution, literally. I show that consciousness is not only of the organism, consciousness is also of the species, and even larger elements than the species. And this is why such a creative leap is possible. There is much evidence in favor of this. The measure of it is that biological evolution moves from simplicity to complexity. When you look at evolution from a consciousness point of view you see this very clearly.

Rahasya: This brings to mind conversations I have had with Bruce Lipton and Peter Russell on the topic of universal consciousness and the possibility that the universe itself is conscious. Indeed, if you research the underlying pattern of evolution you quickly see that it's not only about survival, it's about organisms evolving into higher states of complexity in order to perceive more and, in a sense, become more through forming relationships. What was once a loose collection of cells came together to form multicellular organisms and it seems that the directive of consciousness is to seek higher and higher forms of complexity. So correct me if I'm wrong, but when you say consciousness, are you trying to define it or put it in a box or call it God, or anything specific?

Amit Goswami: This is an important question. I do, but it requires more explanation. I do, because consciousness is the ground of all being, everything is consciousness. Beyond that, definition is impossible because any way we may attempt to define it would limit consciousness. Also, in quantum physics, I find that consciousness has a causal power of choice. The problem with quantum measurement is finding how these quantum possibilities actually become actual events of experience when an observer is witnessing them. This has not been explained and quantum physics will always give you possibilities, never an actual event. So this was solved in my book, *The Self-Aware Universe,* by suggesting that it is consciousness that chooses, and this consciousness is not a dual consciousness because that would also have a paradox of how a nonmaterial consciousness interacts with material. But that's where I made the breakthrough by suggesting that consciousness is the ground of being, or that being is consciousness, or that consciousness is choosing among all possibilities and it does not involve any dualistic interaction involving signals. Interaction is nonlocal, in other words. It's just choosing from its own possibilities, without any signal or energy involvement.

Rahasya: That's a lot to process but it does explain, to some extent, why there seems to be a consciousness involved in the evolution theory. It seems to me that both sides of this argument are so rigid that they are failing to see that each side has a point that complements the other. I can see where this would be a good place to start finding a solution to this debate.

Amit Goswami: Precisely, because with this solution, the idea of consciousness as a causal power comes in and we can call this quantum consciousness, or traditionally, this is what is called God. In other words, God is the causal power of consciousness as the ground of all being. So we can say that God is the causal power exerted in the creative experiences that we have. However, that causal power is usually very limited; we've become conditioned and that conditioning comes from what we call ego.

Rahasya: Yes, and actually that definition works for me, as long as we don't get into "whose God" is causing what. It's the limited definitions of what we're talking about that have needlessly divided us for millennia.

Amit Goswami: There is no "who." God is not a person, God is manifestation itself. We think that God is a superhuman person, but God is not a person. He is not a subject. We can never experience God in a subject/object experience. God is what makes a subject/object experience possible. We can never see God or experience God as separate from ourselves. God is a being but there is no division. In the being there is, however, a force, a choice, that brings in the pertinence of subject and object. The subject and object aspect that our consciousness divides into, has to be illusory. It's only an appearance. This idea was taught in

Eastern thinking; they called it Maya. Maya is the force that causes the perception of separateness of subject and object.

Rahasya: Since our perception and how we see life is dependent to a large degree on what we believe, and what we believe is conditioned by what we perceive, how can we break free when our beliefs seem to have so much power over us?

Amit Goswami: Of course that is the question of creativity. Our beliefs determine our conditioned ego and we don't even realize that our beliefs come from other people. We are so individualistic in this culture, and we don't even realize that all our beliefs are all imposed on us by our society, by teachers, by parents, by the spiritual people that we go to, by our employers, by our newspapers, television, and media. The first sign of breaking down limiting beliefs is that you want to become more creative. And that is the birth of becoming more actual, a more real human being, rather than just being a machine that just propagates other people's learned belief systems. At that point we want to become creative and so we want to get out of the straitjacket of other people's beliefs. This is still difficult because we are comfortable with our conditioning. Once the idea of breaking free from our conditioning comes about, it becomes a burning question. Then things start happening. The creative process comes into play and we start having insights that are beyond our previously held beliefs. Then at some point insights about ourselves come to us, which are very original. We begin to see who we really are, and we find that we are not just converts of these beliefs but we really are beyond these beliefs, but of course when we state our experiences we are generally building beliefs, but at least we can say, "Well, this one is original. I really know that this is the truth because it has come from personal experience."

Rahasya: For me, the importance of personal experience is that it's a form of validation that is not dependent on simply believing something to be true, but knowing it on a deeper level through the actual experience of it. It is important to remember what Bertrand Russell once said: "I would never die for my beliefs because I could be wrong." For me, a belief is nothing more than a steppingstone on the path to knowing, a path that seems to have no end in sight.

Amit Goswami: Yes, validation is an important part of it, because the experience of truth is not an experience in language. It is an experience of what has to be represented in language. So in the representation process, we sometimes bend or misrepresent it. For example, suppose we experience truth, or physical love; how could we ever translate that into mathematics? In the representation process, I may find some mathematics to represent it, but not the entire mathematics that is needed. So I have only an approximation of the truth. So this is why scientific laws are never completely correct; they are only an approximation. Today's belief system is bound to be supplanted by tomorrow's belief system, even though the belief system may be very close to describing reality. In fact, it can never be completely accurate because our language system has imperfection built into it.

Rahasya: To quote George Christoph Lichtenberg, "With most men, unbelief in one thing springs from blind belief in another." It seems to me that since we live in times where we know so much about physical reality that we should take this opportunity to start validating what we believe to be true or not true. This is where science is invaluable. How do you see the relationship between science and religion, or is there a relationship?

Amit Goswami: Well, this relationship can now be better than it has been throughout human history. Historically science has usually kept itself connected with only the explanation of what it calls external. That more or less restricts science to the physical. But then, starting with the 18th and 19th centuries, especially in the 20th and now 21st century, we have extended science to the internal experiences, the science of psychology and biology to some extent. But the main premise of science has still been materialism. That is, there is an underlying material substratum from which our internal experiences arise. This hypothesis, that internal experiences are all material consequences and phenomena, has been a tremendous limitation of current science, of materialist science.

Rahasya: How do we free materialist science from this limitation?

Amit Goswami: Well, for example, the thinking that arises in conjunction with the brain, does not make thinking a brain phenomenon. It could be that the brain, instead, is a representation-making apparatus for thinking, which is happening in consciousness with the help of another aspect called the mind. There are possibilities in consciousness more than just the material possibilities. There are also possibilities of meaning, of feeling, and intuition. We end up with a generalized science, which not only takes the representation-making apparatus, namely matter, but also completes the signal which is being represented in matter. In this way, psychology, biology, and all these are liberated from the material straitjacket and many more phenomena which will not be dealt with in the materialist metaphysics because it is so limiting. So, now we really can open up those sciences and therefore have a more complete picture of the human being and experience.

Rahasya: It seems to me that we are on the cusp of changing our whole worldview and the greatest change is this; we have traditionally held the view that matter creates consciousness, but now, we are starting to see that it's consciousness that creates matter. If this is true, by changing our worldview we could be making or creating the next evolutionary step for mankind where we will actually start perceiving other dimensions?

Amit Goswami: You ask the most important question that is in physics these days. I firmly believe that we are at the crossroad when a new step of evolution from what I call the rational mind to the intuitive mind is going to come about. And we are really moving forward in two ways. One is that although we take pride in the rational mind we're discovering that the mind is also about processing of meaning, it is a fact that meaning processing is still restricted to only an upper echelon of people. It hasn't hit the general population, although we have moved in that direction by creating democracy instead of kings and creating capitalism instead of feudal economies. But still, it is extremely limited and materialism has definitely created a problem in that movement, that whole movement toward democratization has suffered, and now, in the 21st century our main challenge is to get meaning back in our society and in our lives. And I find it extremely curious that certain movements in our economics, certain movements in our politics, are going in that direction. It is very interesting. And of course the groundwork was laid in the 20th century but it's only in the 21st century that we're seeing some of the results.

Rahasya: I think this is becoming the most imperative issue of our time, because living in a society with high technology we have changed the game rules. A hundred years ago if one group of believers wanted to kill 1 million of the "other" believers they would need to gather at least a million of themselves together along with all the logistical necessities to do so, including food, transportation, armament, shelter, et cetera. Now, all we need is one believer with a suitcase bomb to do the same thing. These points are important to consider and get people talking, especially people who have deep conflicting beliefs.

Amit Goswami: So you help people by talking to people and getting them to think. You help to spread the news and you make your livelihood by connecting people. This is an ideal example of what is going to be important for everyone in the new economy.

Rahasya: This is something that I've been interested in lately, because after talking to people like Bruce Lipton and Joe Dispenza and many other scientists, neurologists, and visionaries, I started realizing something about the drawbacks of having a rigid belief system. Here's my question: I would like to get your opinion on something that I am personally interested in. I started thinking about this after reading a quote by Voltaire in which he states, "If they can make you believe in absurdities, they can make you commit atrocities." And after talking with neuroscientists who are doing fMRIs, I realized that there seems to be a neurological reason for this because they have shown that if you are a person who believes in something that lacks any validity and is impossible to validate, that the process of maintaining that belief can cause what they call a limited neurological imprint or limited neuronetworks. When this happens to a person they seem to filter out all new and contradicting information and only let in information that supports their worldview. Of course, religious beliefs are the most obvious because they come under fire with any new information from science. What are your thoughts on this?

Amit Goswami: Well, even without any hypothesis about neural networks, this has been of course an obvious reason why we find fanatics in society. In other words, any good belief systems, like some religions, which are rich with the spiritual insight of some great person like Jesus or Buddha or Krishna or whomever, can eventually bring insight to the followers who are willing to do the work and investigate. But investigation is hard; finding out by investigating takes a long time, long commitment, and those qualities take lifetimes to develop. So what does a lazy or an ordinary person do? Well, they compromise. They start with listening, and listening sounds good, and the followers who are willing to put forth the effort also have the same problem, and the people who teach, they also compromise and they state things in a very simple way which are then taken in a simple way. Good, bad, right, wrong, heaven, hell, and teachings start gradually instilling a sense of fear—if you do this, you will be rewarded, and if you don't do this you will suffer, and so this teaching becomes a dogma, in the sense of this is right, this is wrong, this is good, this is bad, and they become accepted truths without checking them out and eventually become an established dogma or belief system. And once that happens, everything is filtered out through that belief system. All of this happens without even considering the neurological aspect of the situation. But my point is that this is what people have been doing for a long time. So what do we do then? We feel that we should be good, because religion has instilled in us this very dualistic way of being, right, wrong, heaven, hell, those things have become so imprinted in us that we have to then live in such a way that we must support our belief system at all costs. By supporting our belief system we are performing part of our responsibility and at least we can appear to live that way so others might be able to live that way, but we are actually reneging on our own responsibility of being able to live in the way that seems right. That's the first step, and the second step is that we do not live up to our belief systems because of what we call our instincts which are actually brain circuits, and when we can't live up to them we feel so demoralized that we try to lift ourselves up and remoralize ourselves by at least making the belief so firm that we indoctrinate others into it; thinking that if we can convert enough people to this belief system then maybe some people at least will be good and at least I will have done something in favor of what is good.

Rahasya: Thereby getting a reward which means even if it is the right thing to do that you are doing it for the wrong reason. We've become so identified with our belief systems and they have become such an extension of who we feel we are that by bringing others into the fold

we feel that we have somehow left our mark on eternity because it will continue long after we have gone.

Amit Goswami: Yes, we've become so identified and we feel that even though we're not sustaining our belief system maybe somebody will so this propagates the belief. This has been the reason why religions proselytize because they cannot help it. They feel that by proselytizing they are at least helping truth to continue.

Rahasya: I had a spiritual teacher in India named Osho. He was a character and an extreme controversy of terms, but the one thing his teaching did was it tore down my need to pretend I know something without need of evidence, what people call "beliefs."

Amit Goswami: Yes, exactly. What was important is the ability to perform, to create the ability to participate in your own life and this is the first step in doing that.

Rahasya: Yes, for me, he was like a demolition guy. His job was to demolish the building or belief, and unlike so many other teachings, he would not simply build another building/belief for you so you are out in the world in the elements living for the first time with nothing to give you shelter but your own creative being.

Amit Goswami: Yes, exactly. And that is the fundamental component of creativity. First you have to destruct the belief system. You were asking before how do I get out of a belief system. Well, we have to destruct the system. Actually, the role of the teacher has also been misunderstood in a lot of traditions. Traditionally, the teacher is supposed to remove your ignorance. But when you remove ignorance, you start with removing what is causing the ignorance, which is your belief system. So the teacher's job indeed is to first deconstruct your belief system. And then to give you inspiration so you'll go out to create a path to discover what is spirit, what is beauty, what is love, because these things nobody can teach you. So teaching really should be a demolition job.

Rahasya: You need to have your own experience. Sometimes I look at religions like, if I was really hungry, I could go to a religion and they

will give me pictures of food, but those pictures and photos of food aren't really substance, even though the synagogue, temple, church, mosque, or whatever may be beautiful, they have no nourishment. I'm to the point where I want a substantial experience with that which I am seeking, even if it's just a piece of dry bread.

Amit Goswami: Right. But you know, this is the thing, when it comes to spirituality we have undernourished ourselves with the picture of the food. We'd rather talk about the images of God, and to talk about the beliefs we have, the basis of right and wrong, heaven and hell, rather than investigate directly by eating what is being offered and seeing for ourselves whether it really works or not.

Rahasya: Yes, and I can see that in myself. Just to sit down and meditate, to take out 20 minutes a day and meditate seems like such a challenge sometimes. To do the work, the real inner work of sitting quietly and looking inside for my own truth. For more information on Amit Goswami visit www.amitgoswami.org

CHAPTER 36

Dalai Lama and Rick Ray

What Did the Dalai Lama Have to Say about Religion and Science?

The spiritual masters have been telling us for many years to meditate but anyone who's tried to meditate, especially Westerners, knows how difficult that is even if you have time, which you never do . . . or at least it seems so. Our egos have built a world specifically designed to keep us in a trance-induced state that we cannot get out of by traditional means for the simple reason that our traditional means have been created by our egos. One of the impulses behind drug and alcohol addiction is the spirit's attempt to free itself from the mind, if only for a moment. The trouble is that after the high goes away the ego personality is right there to welcome us back to its world, which is a world filled with even more conflict and even more reasons to escape through drugs and alcohol.

The fact that a believer is happier than a skeptic is no more to the point than the fact that a drunken man is happier than a sober one.
George Bernard Shaw (1856-1950)

One of the cleverest things the ego has ever devised is the idea that you can "know" the truth simply because you believe it to be true, and then organize it into a *belief system based on faith* that cannot be questioned or put into doubt for any reason. What this does is to keep us from reasoning, and without the free will to reason we will never get out of this illusion that we are trapped in. After all, if I think I am already awake in a dream, then I will never wake up in that dream. Some very wise people throughout the ages have all told us that we must awake from this life to escape the illusion we find ourselves in; of course we have managed to kill most of them to maintain the illusion. And who is it that kills and persecutes them? The believers with the most power at the time.

To have doubted one's own first principles is the mark of a civilized man.
Oliver Wendell Holmes Jr. (1841-1935)

We must not under any condition give up our right and our ability to reason and to question beliefs that hold us in their grip and limit our understanding of ourselves and our world. I would never want to take your right away to question what I'm saying right now. This, of course, puts me at a disadvantage with believers because they know they're right and will not open themselves to even the possibility that they may be mistaken, whereas I must in all intellectual honesty admit that I do not have all the answers and indeed, I could be wrong about the answers I do have. But I do have a tremendous advantage if I can stay in the arena of rationality, data, facts, and show the obvious human suffering that has a direct connection to beliefs and belief systems.

So what we need to fully understand is that with every physical evolutionary leap there has also been an evolutionary leap of consciousness.

So if it's true that we are about to take another evolutionary step then it seems also true that we are about to take another step in consciousness. This will never come about if we hold on to old traditional beliefs because their very nature is to keep us trapped and therefore never finding any new, unfolding truths.

You might ask why is it that I am so adamant about letting go of our good old traditional beliefs. It's simply this; there is no possible way to get vast groups of people together when each group has the belief that the other group needs to be converted to the first group's belief or it will spend eternity in hell.

I was curious what one of the most respected religious leaders in the world would have to say on this point, so I asked Rick Ray what the Dalai Lama had to say on this issue when Rick interviewed him for his documentary *10 Questions for the Dalai Lama* (2006).

Rahasya: When you asked the Dalai Lama about the conflict in the Middle East, you asked was there hope there, could Buddhism have an impact, and he said he had been to Jerusalem twice and had found a few open-minded Muslims and Jews in the local community and some were making every effort for peace, and then he went on to say that major religions have a potential to create harmony and peace of mind, but because of the politics of people making things so complex, their efforts remain very limited. And he said there's too much negative emotion, like hatred, anger, and frustration. What was his suggestion on what we can do?

Rick Ray: The Dalai Lama said the only thing we can do is to have more picnics and more festivals.

Rahasya: Did he elaborate on that at all?

Rick Ray: No, that was pretty much straightforward. I don't think we cut anything out of that particular monologue he gave. I took a lot from that because I have focused on the Middle East for many years, and I felt that I loved Eastern religions because man and nature are on the same side of the coin and not opposed. But I felt that to be fair I needed to really live in the Middle East and try to understand the Judeo-Christian-Islamic faiths before I dismissed them. I had been raised with them but raised strictly on a very social kind of religious basis and not on the meaning of the faiths, and I really found myself deeply disappointed by what I would call a fundamentalist mentality in these faiths. And yet, I was also very inspired by the mystical mentality of those faiths, too. Like the Sufi tradition or the Kabbalah or certain aspects of the Christian faiths.

Rahasya: Like the Gnostics for instance.

Rick Ray: Yes, exactly. Where they are more prone to say, "I don't know the answer" or to give a kind of a parable of an answer as opposed to saying, "Every word is literally true and if you doubt one word you have to doubt the entire book, and you have to behave according to very rigorous religious dogma and scriptures." These things alienate me and I think the Dalai Lama comes clearly on the side of saying, "We don't know. And let's look at the commonalities and not emphasize these sorts of differences." He says that it is the politicians who often get heard and that it's often the fundamentalists and the fanatics that have the loudest voice and because it's such a sensational voice it's newsworthy, and it's sensational.

Rahasya: And they also have the power behind them, and the power is that they are so sure of themselves, and anybody walking the path and living the truth are filled with doubts, and mystery, and it's all good, but when it comes down to a direct debate, we're at a disadvantage because it looks like we're unsure, we don't know, and we don't. The truth is, neither do they.

Rick Ray: I think you could put the Dalai Lama's position on the side of "don't know" until scientifically investigated. You probably remember from the film that he says that when science disproves something that's in the religious doctrine, you just change the religious doctrine. And that kind of statement really impresses me and makes me feel that he's a one-of-a-kind religious leader.

(I think it's obvious to anyone hearing this that the Dalai Lama is saying that there's nothing we can do because it's obvious that we are not going to get these religious groups together to have a picnic without their blowing each other up in the name of their loving God. If someone such as the Dalai Lama can't come up

with a solution that will work and all of the most enlightened people who have walked the planet these past few millennia have had no success, then I think it's time we look deeper into the problem. It's not this or that religious belief, it's "belief systems" in general that are the problem, and specifically it's the process of believing anything at all that's based on faith and can no longer be addressed rationally.)

The next question was regarding something that was obviously very personal to the Dalai Lama, the subject of the Panchen Lama.[87]

Rahasya: How did you feel when it came up about the disappearance of the Panchen Lama and his family? That surely hits close to home for the Dalai Lama.

Rick Ray: I think I saw something come over him which was not present in the rest of the discussion and that was a true feeling of sadness. And I would say some kind of feeling of heaviness and a feeling of regret that he had played a part in that young boy's destiny. It troubled him that the young boy that was appointed Panchen Lama, through traditional methods, was kidnapped, as the film points out, and has never been seen again along with his family. Whether he's alive or is being held in a detention center where he can't learn the fundamentals of the kind of training that a Panchen Lama would get, the Buddhist wisdoms, his life has basically been shattered, as well as his family's. Within weeks the Chinese appointed their own Panchen Lama using a kind of loophole in the Tibetan Buddhist practice. They normally engage in a search and have very specific parameters, but in certain cases they were able to put

[87] The Panchen Lama bears part of the responsibility for finding the incarnation of the Dalai Lama and vice versa. In the case of the Panchen Lama, the religious procedures traditionally involve a final selection process by the Dalai Lama. There is a controversy about who is the true present (11th) incarnation of the Panchen Lama: The People's Republic of China asserts it is Qoigyijabu, while the Tibetan Government in Exile maintains it is Gedhun Choekyi Nyima, who was arrested at the age of six years by the Chinese in 1995. He then became the world's youngest political prisoner. This is but one more way that governments use the beliefs of people to control them. If it comes about that the Chinese government installs its own Panchen Lama, it will appoint the new Dalai Lama, who will of course adhere to the rules and conditions of the state since he will actually be a politician in disguise.

names in an urn and draw them at random. I don't know if that was in times of drought or when travel was very difficult, but the Chinese used that, called the Golden Urn exception, as an excuse to choose from a series of monks that they had chosen themselves and handpicked.

I would say that their strategy is also very clear, which is a long-term plan to wait out the Dalai Lama's lifetime rather than taking any action against one of the world's most popular and beloved personalities. Just wait until he passes away and then just quietly take over the appointing of these leaders. So that will be done quietly and probably without a lot of press and attention, so we might be seeing the last true Dalai Lama.

(I thought it was curious also what the Dalai Lama had to say about population control and even more interesting what he did not say; I think this has a lot to do with the fact that he has become such a political figure in the world and has to do a balancing act between that which he knows to be true and that which can be said without causing harm to large populations of people.)

Rahasya: The Dalai Lama says that we should limit population growth because of limited resources, and I agree, but we inevitably run up against the problem of who is going to limit themselves, and it's usually the most informed and intelligent people who do this. He says that quality is more important than quantity. I also agree. But it seems we need a system of enforcing a limited population growth, which also brings up another obvious problem of who gets to decide which population gets limited. I'm just wondering, did he elaborate on that at all?

Rick Ray: Unfortunately, he didn't elaborate much on that particular question, and I wish he had. And I can't really speak for him as to what he'd recommend. I can say that he thinks he has to be careful when he speaks about these things because there are other religious traditions out there that he respects, or even embraces, that might take a different approach to population control, birth control, and things like that. We do know that he is an environmentalist at heart, and he would like to see population regulated worldwide somehow. He also talks a lot about discipline, especially self-discipline. So, I would imagine his answer would probably have more to do with some kind of self-discipline toward reproduction than some kind of external law from outside that might or might not be effective. This led into a discussion of the gigantic population of India, and its definite impact on the quality of life there.

Rahasya: What were the Dalai Lama's thoughts on peaceful resistance and how the world seems to set aside what China has done in favor of commerce? A good example of course is what Google, Yahoo, and Microsoft have done to censor their websites and search engines and how Western society buys billions of dollars worth of products, some of which are dangerous and even toxic to our children. What did he say about that?

Rick Ray: One of the questions that we do ask in the movie is: "At what point does peaceful resistance not serve you; at what point do you take action against an enemy or perceived enemy, if your life is at stake, or they're directly threatening your nation or your children or yourself?" And his answer, I thought, was very surprising. It's a practical answer, typically, because he's such a practical individual, but he said, "You know, if you're just a pacifist, if you think as a pacifist all the time, then if somebody comes to kill you you're not going to live to be a pacifist another day. So if there's a direct threat, if someone's coming to cause you harm, then it's okay to just hit back," and then he makes a gesture with his fist. But this is just for self-defense, and it begs the question of when do you really consider it self-defense? How do you justify on a larger scale the question of what self-defense is? When a country is vaguely threatening you, is it self-defense to go and hit them? I don't think he would advocate for that. But on a personal level, he says, "Defend yourself to live another day."

None of these questions are easy or simple, but again, what comes to the foreground with a little thought is that most of these complexities and absurdities in our world are caused by beliefs that have been given special dispensation from inquiry and no longer, if ever, need validation. A good example of this is that in China if you want to reincarnate you need to get a Reincarnation Permit. If this were the only example of how religious people can be herded around it would be enough but as you know by now, there are many such absurdities in our world.

CHAPTER 37

From Creationism to Evolution to Intelligent Design to "What the Bleep Really Happened?"

All truth passes through three stages.
First, it is ridiculed. Second, it is violently opposed.
Third, it is accepted as being self-evident.
Arthur Schopenhauer (1788-1860)

I had the most apprehension about writing this chapter in particular, not because the information is without merit, but because it is so far beyond what the average person would or could deem credible. That's why I preface it with information from what the "average" person would call credible sources. I think that with all the evidence that is available, testimonies from high government officials on their deathbeds, photos, and videos, we need to move past the conversation of whether or not there are advanced beings on other planets and move on to the next question, which is, "What do they want?"

If you recall, I have already mentioned that I don't want you to simply "believe" what is in this book; I want you to weigh the evidence and come to your own conclusions by thinking for yourself. The best way to keep someone from thinking is to implant a limiting belief that claims to be the ultimate and complete truth. It's important to remember the pattern of secrecy that's been the standard operating procedure with most of our political and religious organizations throughout history. Don't assume for a moment that things have changed. And remember that my purpose is not necessarily to give you something else to believe in; it's to help make clear that the beliefs we do have need to be reexamined in light of recent information and more advanced scientific research.

I want to introduce you to Monsignor Corrado Balducci. He was born in 1923 in Italy and died in 2008. He was a Roman Catholic theologian of the Vatican Curia. He was also an exorcist for the Archdiocese of Rome and a prelate of the Congregation for the Evangelization of Peoples and the Society for the Propagation of the Faith. Monsignor Balducci graduated from the Pontificia Accademia Ecclesiastica in 1954. This is the training institute for priests who will become papal ambassadors. But probably most important of all, he was considered a Vatican insider and a close friend of the last two popes. In short, if you're looking for credentials, Monsignor Balducci's are impeccable. He says his interest in extraterrestrials and UFOs is personal but we can be assured that everything he says is sanctioned by the Catholic Church and the pope because he has never received any condemnation for his statements.

He met with Zecharia Sitchin for the purpose of discussing Mr. Sitchin's work on the Sumerian clay tablets and some of Sitchin's conclusions about human history. Indeed, some might consider the meeting a historical event; after all (in Zecharia Sitchin's words), "It isn't every day that a high representative of the Vatican and a descendant of Abraham get together to discuss extraterrestrials, UFOs, and the creation of man."

After much talking on the subject they came to some common conclusions, three of which are:

1. Extraterrestrials can and do exist on other planets.
2. They could very well be more advanced than us.
3. And the most important of all, man could have been fashioned from a preexisting sentient being that evolved naturally.

Monsignor Balducci also said, "Atmospheric phenomenon and optical illusions aside, there are simply too many eyewitness sightings, so there has to be something to it." He also said, "Such witnessing cannot be dismissed by a Catholic theologian because it is an important aspect of transmitting truth and in the case of the Christian religion we are talking about a Divine Revelation in which witnessing is crucial to the credibility of our faith." He quotes Cardinal Nicolo Cusano (1401-1464), who said that there is not a single star in the sky about which we can rule out the existence of life, even if different from ours.

Balducci also brought up the point that the Church recognizes that there are angels who have no bodies and are pure spirit and there are humans who have bodies and also have spirit. He concluded that the distance between these two creations is great and there must be intermediary beings, which also means that they, extraterrestrials, would be more spiritually advanced than us. He also points out that they are not to be considered as evil. He said that despite what people think, the Church would be able to reconcile their existence.

At one point during their conversation Zecharia Sitchin quoted from Sumerian texts, which say the Anunnaki, "those who came from heaven to earth," genetically improved the existing beings on earth to create the intelligent being, the same being the Bible calls Adam (I might point out that Adam means "Earthling" as it is used here). Sitchin asked if Balducci saw a conflict with this. The monsignor said that more than anything else, Sitchin's approach was based on physical evidence; it concerns itself with matter, not spirit. He went on to say, *"This is an important distinction, because if this distinction is made, I can bring up the view of the great theologian professor Father Marakoff, who is still alive and who is greatly respected in the Church.* He formulated the hypothesis that when God created man and put the soul into him, perhaps what is meant is not that man was created from mud or lime but from something preexisting, even from a sentient being capable of feeling and perception. *So the idea of taking a pre-man or hominid in creating someone who is aware of himself is something that Christianity is coming around to. The key is this distinction between the material body and the soul granted by God."*

I might also point out that the Vatican has two observatories, one in Italy in the mountains southeast of Rome, and the other one in Arizona. It makes one wonder why the Vatican would have its own astronomical observatory.[88] Pope John II may have answered this question at the Holy Year 2000 celebrations when he offered his views on religion and science by stating, *"Scientific research is a genuine way to arrive at the source of all truth revealed to us in the Scriptures."* Father José Funes, a Jesuit priest with a degree in astrophysics and one of the luminaries of the Vatican Observatory, stated, *"extraterrestrials exist and they are our brothers."* The focus of the observatory is made clear by the observatories director, Father George V. Coyne, in one

---

88 This is almost laughable in light of how the Church persecuted Galileo, with the pope at that time refusing to even look through his telescope.

of his public writings and lectures called, "The Evolution of Intelligent Life on Earth and Possibly Elsewhere."

As far as what other theologians have to say on this matter, let me quote a few:

1. Cardinal Nikolaus Cusanus (1401-1464) stated, *"We are not authorized to exclude that on another star, beings exist which are completely different from us."*
2. The Jesuit priest and astronomer Angelo Secchi (1818-1872) wrote, *"It is absurd to claim that the worlds surrounding us are large, uninhabited worlds and that the meaning of the universe lies just in our small, inhabited planet."*
3. The famous Dominican preacher Jacques-Marie-Louis Monsabre (1827-1907) referred to the principle *"Natura non facit saltus"* (Latin for "Nature does not make [sudden] jumps") when he claimed that other intelligent beings besides men and angels exist.
4. The English Jesuit priest Herbert Thurston (1856-1939) wrote, *"Who can claim that there are no other intelligent beings besides angels, demons and men in the Universe? I do not intend to confirm the possibility I indicated in my question as a fact, but I ask: Who can be sure about it."*
5. Then there is the great stigmatized Capuchin Padre Pio, who was beatified by Pope John Paul II on May 2, 1999, and canonized on June 16, 2002. From Saint Pio, the following dialogue is documented and officially published by the Capuchin Order:

Question: Padre, some claim that there are creatures of God on other planets, too.

Answer: "What else? Do you think they don't exist and that God's omnipotence is limited to this small planet Earth? What else? Do you think there are no other beings who love the Lord?"

Question: Padre, I think the Earth is nothing compared to other planets and stars.

Answer: "Exactly! Yes, and we Earthlings are nothing, too. The Lord certainly did not limit His glory to this small Earth. On other planets other beings exist who did not sin and fall as we did."

(Don Nello Castello: *Cosi parlo P.Pio,* Vicenza 1974)

The noted author and Vatican expert Father Malachi Martin suggests that the Vatican is concerned that it will have a major doctrinal updating situation on its hands when extraterrestrial contact becomes authoritatively announced by world governments within the next several years.

I could actually write a book on well-respected religious figures who also hold this type of open-minded view on extraterrestrials and our questionable beginnings on this planet. It's mostly the uninformed believers who have alienated themselves from further learning because it implies "a loss of faith." This makes it almost impossible for them to assimilate conflicting information such as this.

We should also take a quick look at the possibility that we have had contact in our past and that it has been recorded according to the best understanding that people had at the time. One of the many, and possibly one of the best, sources for this turns out to be the Bible itself. It may seem as if I'm going off-topic; however, stay with me on this subject because it is important in regard to information that will be brought out later. Remember, we are questioning our answers in light of new information, and this new information sheds light on what may have influenced the early writers of the Bible and other sacred texts such as the Gita, Ramayana, and the Mahabharata.

Straight from the Bible

What I include here is straight out of the Bible. You need to keep in mind that we are reading the firsthand accounts of people who lived in a time when there was no technology, no flying machines, and the scientific knowledge of such things was not in the consciousness of the people living then. So it only makes sense that when they would see something completely out of their experience that they would identify it in the terms that would be applicable for them and the society at that time.

For instance, what was to be found in the sky in that time? There were birds and there were clouds and that was it, so their terminology for anything flying in the sky would tend to be labeled as such. For example, when we hear the term "flying saucer" we never give it a second thought. It never enters our minds that it might be an actual saucer from someone's cupboard. Even in modern times we know that when the astronauts said, "The eagle has landed," they did not mean an eagle landed literally. There are more than 70 verses in the Bible that describe clouds as vehicles, making them the most common descriptions of flying vehicles of the Elohim or what we would call flying saucers today.[89]

If you consider the size of the objects reported then, and now, it's not surprising that they would be called "clouds." When you add to this the further descriptions that the clouds were "glowing, moving rapidly, leaving a pillar of smoke or fire," and other anomalous behaviors it leaves little doubt that they were much more than clouds. This gets really interesting when you remember that Jesus supposedly took off into a cloud and will someday appear again in the same manner.

What is clear is that the writers of the Bible were obviously not talking about clouds. Let's look at a few passages in the Bible just to give you an idea.[90] And keep in mind the lack of technological language at the time. It's also easy to understand why these encounters would be seen as encounters with God and not as beings from another planet.

Jeremiah

4:13: Behold, he shall come up as *clouds*, and his *chariots* shall be as a whirlwind: his horses are swifter than eagles.

Ezekiel

1:4: And I looked, and, behold, a whirlwind came out of the north, a great *cloud*, and a fire infolding itself, and a brightness was about it, and out of the midst thereof as the colour of amber, out of the midst of the fire. 5: Also out of the midst thereof came the likeness of *four living creatures.* And this was their appearance; *they had the likeness of a man.*

(It's written that a great thundercloud flying craft came out of the north. The center was illuminated and like polished metal and shaped like a wheel that was spinning. In today's modern era this would surely be written as an encounter with an unidentified flying object and extraterrestrials.)

89 The word "`*elohim*" is a word that designates a "classification" of beings or entities and is considered plural. It is like other classification words,for instance, mankind, animals, birds, and so on. It took on the meaning of "God" or "Gods" much later.

90 From www.bibleufo.com

Nahum

1:3: The Lord is slow to anger, and great in power, and will not at all acquit the wicked: the Lord hath his way in the whirlwind and in the storm, and the *clouds* are the dust of his feet.

(Could the whirlwind of Y'hovah be his craft?)[91]

Exodus

16:10: And it came to pass, as Aaron spake unto the whole congregation of the children of Israel, that they looked toward the wilderness, and, behold, the glory of the Lord appeared in the *cloud.*

24:18: And Moses went into the midst of the *cloud*, and took him up into the mount: and Moses was in the mount forty days and forty nights.

(In this instance, Moses was actually taken up, which is further evidence that it was not a cloud into which he was taken up but something much more substantial.)

34:5: And the Lord descended in the *cloud*, and stood with him there, and proclaimed the name of the Lord.

40:34: Then a *cloud* covered the tent of the congregation, and the glory of the Lord filled the tabernacle. 35: And Moses was not able to enter into the tent of the congregation, because the *cloud* abode thereon, and the glory of the Lord filled the tabernacle. 36: And when the *cloud* was taken up from over the tabernacle, the children of Israel went onward in all their journeys. 37: But if the *cloud* were not taken up, then they journeyed not till the day that it was taken up. 38: For the *cloud* of the Lord was upon the tabernacle by day, and fire was on it by night, in the sight of all the house of Israel, throughout all their journeys.

Numbers

9:16: So it was always: the *cloud* covered it by day, and the appearance of fire by night. 17: And when the *cloud* was taken up from the tabernacle, then after that the children of Israel journeyed: and in the place where the *cloud* abode, there the children of Israel pitched their tents.

(It's clear that the "cloud" was guiding the children of Israel and showing them where to camp; the vehicle hovers for days)

---

[91] Y'hovah: God's real name, pronounced "YeHoVaH" (God loves you, you are not alone, you are a child of Y'hovah).

9:18: At the commandment of the Lord the children of Israel journeyed, and at the commandment of the Lord they pitched: as long as the *cloud* abode upon the tabernacle they rested in their tents. 19: And when the *cloud* tarried long upon the tabernacle many days, then the children of Israel kept the charge of the Lord, and journeyed not.

(Now the "cloud" seems to be letting them know when to move.)

9:22: Or whether it were two days, or a month, or a year, that the *cloud* tarried upon the tabernacle, remaining hereon, the children of Israel abode in their tents, and journeyed not: but when it was taken up, they journeyed.

(The word "upon" as used here is from the word "al" *and means "above" or "over." The craft would hover over their encampment or tabernacle for weeks and sometimes many months and they would move only when the craft would move.)*

10:11: And it came to pass on the twentieth day of the second month, in the second year, that the *cloud* was taken up from off the tabernacle of the testimony. 12: And the children of Israel took their journeys out of the wilderness of Sinai; and the cloud rested in the wilderness of Paran.

(This pattern of moving from place to place goes on for 40 years, and all the while they are following an ascending, hovering, and descending craft.)

10:34: And the *cloud* of the Lord was upon them by day, when they went out of the camp.

Deuteronomy

1:33: Who went in the way before you, to search you out a place to pitch your tents in, in fire by night, to shew you by what way ye should go, and in a *cloud* by day.

5:22: These words the Lord spake unto all your assembly in the mount out of the midst of the fire, of the *cloud*, and of the thick darkness, with a great voice: and he added no more.

1 Kings

18:44: And it came to pass at the seventh time, that he said, Behold, there ariseth a little *cloud* out of the sea, like a man's hand. And he said, Go up, say unto Ahab, Prepare thy chariot, and get thee down that the rain stop thee not.

(Today we have many sightings at sea.)

We need to clarify here that it was probably not God who was flying around in these crafts. If we keep that in mind, then we can take a fresh look at some other parts of the Bible. Remember that in those days, the only things that would be considered as lights were stars, the sun, a lamp, and a campfire; *anything else, such as artificial light, would be considered a miracle.*

Genesis

15:17: And it came to pass, that, when the sun went down, and it was dark, behold a smoking furnace, and a burning lamp that passed between those pieces.

(This was the first mention of a flying object associated with the Elohim.)

19:24: Then the Lord rained upon Sodom and upon Gomorrah brimstone and fire from the Lord out of heaven.

(If this was not the crazed and vengeful God of the Bible, then it makes one wonder what was being dropped on Sodom and Gomorrah.)

Exodus

3:2: And the angel of the Lord appeared unto him in a flame of fire out of the midst of a bush: and he looked, and, behold, the bush burned with fire, and the bush was not consumed. 3: And Moses said, I will now turn aside, and see this great sight, why the bush is not burnt. 4: And when the Lord saw that he turned aside to see, God called unto him out of the midst of the bush, and said, Moses, Moses. And he said, "Here am I."

(Keep in mind that until almost 2,000 years after the last book of the Bible was written there were no forms of artificial light and any such light would be a miracle by their standards.)

9:23: And Moses stretched forth his rod toward heaven: and the Lord sent thunder and hail, and the fire ran along upon the ground; and the Lord rained hail upon the land of Egypt. 24: So there was hail, and fire mingled with the hail, very grievous, such as there was none like it in all the land of Egypt since it became a nation.

(It's obvious that something was happening that had never happened before and beyond anything they understood.)

13:21: And the Lord went before them by day in a pillar of a cloud, to lead them the way; and by night in a pillar of fire, to give them light; to go by day and night: 22: He took not away the pillar of the cloud by day, nor the pillar of fire by night, from before the people.

(The word for "pillar" is translated from "ammuwd" *and means a column as standing; also a stand, i.e., platform.)*

14:24: And it came to pass, that in the morning watch the Lord looked unto the host of the Egyptians through the pillar of fire and of the cloud, and troubled the host of the Egyptians.

19:18: And Mount Sinai was altogether on a smoke, because the Lord descended upon it in fire: and the smoke thereof ascended as the smoke of a furnace, and the whole mount quaked greatly. 19: And when the voice of the trumpet sounded long, and waxed louder and louder, Moses spake, and God answered him by a voice.

20:18: And all the people saw the thunderings, and the lightnings, and the noise of the trumpet and the mountain smoking: and when the people saw it, they removed, and stood afar of.

24:17: And the sight of the glory of the Lord was like devouring fire on the top of the mount in the eyes of the children of Israel.

40:34: Then a cloud covered the tent of the congregation, and the glory of the Lord filled the tabernacle. 35: And Moses was not able to enter into the tent of the congregation, because the cloud abode thereon, and the glory of the Lord filled the tabernacle. 36: And when the cloud was taken up from over the tabernacle, the children of Israel went onward in all their journeys. 37: But if the cloud were not taken up, then they journeyed not till the day that it was taken up. 38: For the cloud of the Lord was upon the tabernacle by day, and fire was on it by night, in the sight of all the house of Israel, throughout all their journeys.

Leviticus

9:24: And there came a fire out from before the Lord, and consumed upon the altar the burnt offering and the fat: which when all the people saw, they shouted, and fell on their faces.

(I have refrained from offering too much interpretation on purpose. It's important that you make a shift in how you see the world and how you interpret what you hear and read. We rely too much on others to do this for us. With a little honest thought you will find that these statements from the Bible have nothing to do with a God who flies around messing with peoples' lives; this is something else entirely.)

Isaiah

66:15: For, behold, the Lord will come with fire, and with his chariots like a whirlwind, to render his anger with fury, and his rebuke with flames of fire.

2 Kings *(Spinning Chariots of Fire Take Elijah)*

2:11: And it came to pass, as they still went on, and talked, that, behold, there appeared a chariot of fire, and horses of fire, and parted them both asunder; and Elijah went up by a whirlwind into heaven. 12: And Elisha saw it, and he cried, My father, my father, the chariot of Israel, and the horsemen thereof. And he saw him no more: and he took hold of his own clothes, and rent them in two pieces.

(There are many instances in the Bible when God spoke to the people from a whirlwind. A note of interest is the word that was translated into "horses" actually means "rapid movement or flight.")

Many more parts of the Bible and other ancient scriptures lead a thinking person to see the possibility that we have been visited before by beings from somewhere else. To think otherwise would require one to believe that clouds did some pretty strange things in ancient times that they no longer do or that whole civilizations were part of a grand hoax that spanned many thousands of years.

Mahabharata and the Ramayana Ancient Indian Texts

At the risk of putting you into informational overload, I think it's important to point out that almost every culture that dates back into antiquity has similar stories about UFOs and extraterrestrials, although they may not call them by our familiar terms. Take the great national epic from India, the Mahabharata. It dates from between the 8th century BC and the 4th century AD and is a collection of about 80,000 papers or couplets. The Mahabharata tells the story of the long and brutal war between the Kauravas and the Pandavas. Apparently the "Gods" came to the conclusion that there needed to be some population control because there were far too many human beings, which makes one wonder what those Gods would think today with almost 7 billion people living on the planet.

It is in the Mahabharata that we hear about Bhima, who "flew in his *vimana* (what we would call a flying saucer) on an enormous ray which was as brilliant as the sun and made a noise like the thunder of a storm." This is where we hear of Arjuna, who was taken off to heaven (Indra) up in the clouds in a flying vehicle that made a noise like thunder. Arjuna is very articulate and leaves little doubt as to whether he is speaking symbolically or not. He speaks of seeing other flying objects; some of them have crashed, some are stationary, and some are flying.

The Mahabharata also speaks of the terrible weapons in the possession of these beings, weapons that seem eerily like our atomic weapons of today. They speak of the unthinkable destruction caused by these weapons, which might lead one to speculate upon the synchronicity of the influx of UFO sightings right after we started experimenting with atomic weaponry. In Erich Von Daniken's book *According to the Evidence*, he describes evidence of such terrible weapons that were in the possession of Indian Gods in Book 8, *Musala Parva:* "*The unknown weapon is radiant lightning, a devastating messenger of death, which turned all the relations of Vrishni and Andhaka to ashes. Their calcined bodies were unrecognizable. Those who escaped lost their hair and nails. Crockery broke without cause; birds turned white. In a very short time food was poisonous. The lightning subsided and became fine ash.*"

In the Ramayana, often considered the second great Indian epic after the Mahabharata, one can find stories of *vimanas* that flew great distances and heights with a "great propulsive wind" behind them. The Ramayana also speaks of powerful weapons. In the tale of "Rama and Sita," the king promises Rama that if he succeeds in bending an immensely powerful bow, he can have his daughter, Sita, the princess born from the earth (daughters of men in the Bible). "*Straight away the King gave the orders: five thousand well-built men strained to drag the eight-wheeled iron cart which contained the divine weapon. Try, commanded the devout king, and effortlessly Rama seized the bow and drew it. He bent it to such an extent that he broke it and the sound that the string made in breaking struck fear in everyone who witnessed it. My daughter, said the king, will be the prize won by your strength.*"

One of Rama's comrades in the Ramayana called "The King of the Apes" was the pilot of a tremendously powerful flying machine that lifted off vertically and didn't need a runway and shook the ground and the trees. In fact, there are many instances where flying vehicles are mentioned in ancient texts, which implies there were ancient astronauts.

With the evidence we have, it is at least reasonable to deduce that our ancient ancestors were doing their best to describe events that were far beyond their linguistic capabilities to describe because of the lack of technical terms. If these events were going on today we would describe it as an occupation or an all-out invasion of our planet. In the same way we have occupied and invaded foreign lands, it's possible that all of our visitors did not have the best of intentions while here. Of course I do not want to be an alarmist, because the way I see it, if they, whoever they may be, had bad intentions for us now, there's probably not too much we could do about it anyway.

Consider this: Look how far we have advanced in 100 years. Imagine for a moment that we had just one modern-day aircraft carrier back in World War II. With that one ship we could have completely dominated the planet, and this is after less than 100 years of technological evolution. Now, think of how advanced a civilization might be after 1,000 years, or 10,000 years, or even millions of years? Its advancement would not only go beyond what we have ever thought or imagined, it would go beyond what we "can" imagine. So if it is true that other beings are visiting us and will someday land on the White House lawn, then we need to stay out of the arena of conflict and fear. The approach espoused in statements such as this one by evangelical Christian Pat Robertson: *"Anyone speaking of aliens should be taken away and stoned to death because they are part of Satan,"* would definitely not be the best approach to take with an advanced technological society.

CHAPTER 38

Swami Kriyananda and Rahasya

Belief and Experience
An Interview with Swami Kriyananda

While I was doing the research on the Mahabharata and the Ramayana it occurred to me that it would be interesting to get some feedback from someone who was well versed in those scriptures and Eastern thought. There is probably no one better than Swami Kriyananda, who was one of the first disciples of Paramahansa Yogananda in the United States in 1948. Swami Kriyananda is more than 80 years old now and still teaches in the United States and India. This interview took place at the Ananda Retreat just outside of Nevada City, California. We sat in his living room overlooking the 800-acre retreat. He has dedicated his life to the experience of that which people refer to as God, experience being the key word here. For him, belief is . . . well, I'll let him tell you.

Rahasya: Whether it's governments, corporations, or religions it seems that every organization that is ego based and built on old paradigms of competition and secrecy are breaking down. In regard to religion what do you think the difference is between religion and spirituality?

Swami Kriyananda: Religion is organized, and spirituality is what the individual feels in his relationship with truth and with God. And although spirituality may be expressed in a religion, many people are spiritual and never go to church. They are not religious in the sense that they practice a certain type of discipline.

Rahasya: I can only speak for myself but on a personal level the only thing meaningful I have ever learned of spirituality or what some would call God, has been experiential and, actually, would be impossible to transmit that to another person through traditional communication.

Swami Kriyananda: Yes, you cannot go further than that. You can accept that it may be true what somebody else says, but until you have experienced it, it is not dynamic to your own awareness.

Rahasya: I was talking with my daughter once about something and she asked, "What do you mean?" and I said, "Well, it would be like if I eat a piece of fruit, I can taste it, I can feel it, I can be nourished from it, and I can tell you about it, but . . ."

Swami Kriyananda: Until you've tasted it, you can't really know it.

Rahasya: Yes, there is spiritual nourishment that comes only through direct contact and experience of any truth. This is what I think gets lost in a lot of belief systems built around rigid dogmas.

Swami Kriyananda: Absolutely. Spirituality means seeking experience. Not just belief.

Rahasya: With what's going on the world today, I think it is really important to have dialogue with people that you don't agree with, and for curiosity I called up Amnesty International, and they told me that today there are 15 religious wars going on in the world. And these wars are usually based upon collective beliefs. Of course a lot of these wars are people with agendas hiding under the umbrella of a particular religion.

Swami Kriyananda: Abstract beliefs.

Rahasya: Yes, I'm talking about the kind of beliefs where you're pretending that you know something that really you don't know and probably have no way of knowing.

Swami Kriyananda: Quite right.

Rahasya: What are your thoughts on these kinds of conflicts?

Swami Kriyananda: Well, I think they are just what you say. They are ego based; it's a mistake to think that God has conflict with anything. He is everything. So when you are close to God, how can you be in conflict with anybody? Conflict comes from ego, and from thinking, "I am right and you are wrong." If I can reach the point where I understand that what is right for me may be different than what is right for you, that would be a good step. But most people don't reach that point, and so they fight about it.

Rahasya: So, if you were to, and I have a feeling you would not, but if you were to attempt to define God, what would your definition be? Or do you think it's a definable thing?

Swami Kriyananda: You can't define God, any more than you can define consciousness. God is conscious bliss. How can you define either of those terms? So defining is a mistake the intellect imposes, thinking you'll understand when you define. You don't. You understand when you experience.

Rahasya: There was a saying by Voltaire, and the saying goes like this: "If they can make you believe in absurdities, they can make you commit atrocities."

Swami Kriyananda: That's a very good saying.

Rahasya: Yes. And it started me thinking one time, "Why is that true?" And looking into it and really researching it, if I can make somebody believe something completely absurd, like there's a 10-ton gold brick buried in their backyard I can start controlling that person because he has detached himself from reality to some degree and I can use that detachment to manipulate him or her.

Swami Kriyananda: If you have reached the point where you have convinced him against his reason, and against his experience, you have brought him to the point where he will be able to believe anything and do anything. That is a great line.

Rahasya: That's the reason it's so important to validate as much as possible.

Swami Kriyananda: Absolutely.

Rahasya: I have beliefs, the belief takes me to a small step, and then I look around for validation and then take the next step with what I believe may be true.

Swami Kriyananda: Belief is hypothesis. You have to experiment, which means experience and that experience will tell you whether it is true or not. People believed the world was flat, but that did not make it so. They have many beliefs that are not true. But if you try to put yourself in a position to try to seek provable beliefs, you cannot believe, you cannot know whether the world is flat or round except by going around it, but you can know whether being kind to people is more satisfying than being unkind; whether love is more satisfying than hate. And all these qualities that we think of as moral or virtuous and so on, every one of them should be tested. It is not enough to be following a belief and virtue or morality because your society or your parents or your teachers tell you. In our schools, we emphasize giving children a chance to find out from their own experience. This is the way to begin with seeking God, too. You cannot know whether there is an infinite consciousness, but you can know whether something makes you happier in yourself. And this is a good start.

Rahasya: Yes, I like what Aristotle was talking about, morality and everything and he said, "Just as a matter of fact, I like to be around people that like me. So it only makes sense that I treat them well." It can be as basic as that.

Swami Kriyananda: Yes, sure it can. And I think we need to begin with basics. One question, and it is a basic one. How do you know that you exist? Descartes solved it, according to him, by saying, "I think, therefore I am." But that doesn't prove anything, because when you are *really* aware, you don't think.

Rahasya: Yes.

Swami Kriyananda: And the more you think, the less aware you really are. So that doesn't do it. You can't know by reason whether you exist or don't exist. It's an intuition. And this intuition that I exist is the beginning of human awareness. And the beginning of human seeking. I know that I exist, I know that I am conscious. How can I become more happily aware of this stage? How can I become more conscious? This is the beginning of all seeking. Not whether the right teacher is Jesus Christ or Buddha or anyone else, but whether what they taught will help you become what you want to be. Everybody wants to be happy and nobody wants to be unhappy; everybody wants to be calm, and nobody likes to be agitated. It always comes down to what expands your self-awareness or what limits it.

Rahasya: For about the past 80 years or so, we have been finding Sumerian clay tablets that were from Mesopotamia and to even older civilizations. And on these clay tablets, there are about 30,000 of them, one of them is the Epic of Creation, which is almost a direct translation on the story of Genesis in the Bible. Since it predates the Biblical account by centuries, perhaps even millennia, it's become

obvious to Biblical scholars that the Old Testament, especially the Torah, the first five books are copied from older Sumerian clay tablets.

Swami Kriyananda: That's quite possible.

Rahasya: What the clay tablets say is that the Anunnakis came here from another planet called Marduk, which has a 3,600-year orbit around our sun. Zecharia Sitchin and a few other people have written about this extensively. Supposedly they have been visiting our planet for the past 450,000 years. It's written that they genetically altered the sentient beings already here, the **homo erectus,** ***and basically brought us up to the level of what we would refer to as*** **homo sapiens.** ***This of course brought us to self-awareness and we came to mistakenly call Anunnakis Gods. As a matter of fact, the word*** **nephilim,** ***which is translated as "giants" in the Bible, means "those who came down from above." So all this has to make a person wonder when we read any of our ancient texts that have been translated.***

Swami Kriyananda: You know one problem you get with translators is that they can't get away from their own ideas. And so they will translate according to the society and beliefs and of course words and concepts also change their meaning over time. So what do I believe about all this?

Rahasya: Yes.

Swami Kriyananda: Well, I believe that there are other life forms, that space is full of life. There certainly have been visitors from outer space. As far as this explaining the link between *homo erectus* to *homo sapiens,* Yogananda said they [anthropologists] will never find a missing link. He said man was created as an act of God. Now whether that act of God came from another planet or not, I don't know, which I think is the point you are trying to bring out. In my opinion, it had to begin somewhere. I'm not at all averse to the idea that people have come here from other planets and intervened into our genetics. I can think of the Chinese and the Japanese; their facial structures are so different, you can't help but wonder if we all came from the same racial beginnings. But the interesting thing is that all the definitions, all the descriptions of aliens, so to speak, they all look more or less like human beings, and to me this is a very good answer to Darwin, that there was something pushing it toward that direction. Because otherwise you could think of them with tentacles, or whatever. It doesn't seem to be so. That thing at Roswell, where they found some actual cadavers that were made differently. They had some sort of vegetable beginning, but they looked like people, and I would be inclined to think that "yes, there's life everywhere, and intelligent life will have to express itself as something similar to the human body." It's the most natural expression of that which can evolve toward spirit. The spine, the rising of the energy in the spine, this has to be the essence of advanced life everywhere. So where does this take us?

Rahasya: Indeed, where does this take us?

CHAPTER 39

So What Really Happened? How Did We Get Here? Where Did We Come From?

So many questions and so very few answers, but we do have some interesting facts to go on that seem to point us in a direction that will surprise creationists and evolutionists alike.[92] You may recall that in the interview with Amit Goswami we discussed some of the deficiencies of both the Darwinian evolution theory and the Intelligent Design theory. We're going to take a fresh look at these theories from an entirely different perspective with new information. I will warn you ahead of time that this might be difficult to swallow but as it was once said, "*All great truths were blasphemies when first spoken.*" I'm including this information in my book because, one, it is backed up by hard evidence as found in the Sumerian Tablets and other archaeological discoveries, and two, this could be an important key to understanding why it is that creationists seem to have a point and evolutionists also seem to have a point and at the same time they both seem to be missing an important piece. This would also explain the "missing link" that we hear so much about.

But I must bring something up; if you are a creationist and expect this to be comprehensible at all you will need to accept that the world is probably older than 6,000 years, much, much older.[93] And if you are an evolutionist, you will need to accept the fact that there are missing pieces in your theory;

---

92 *Genesis Revisited* by Zecharia Sitchin, 1990

93 In 1648 Archbishop James Ussher calculated the beginning of the world by using the information in the Genesis genealogy section. In this work, he

no matter how much evidence you have to back it up, there are still some issues that simply do not make sense.

Genesis Revisited

Science has proof without any certainty.
Creationists have certainty without any proof.
Ashley Montague (1905-1999)

It's important to remember the goal and purpose of this book; its purpose is not to convince you, thereby supplanting one belief with another. The purpose is to dislodge the need to believe altogether and to get you to think for yourself, and if need be, to be comfortable with "not knowing" while searching for the truth. One of the ways to do this is to bring yourself to a new perspective through reason and logic and seeing the absolute impossibility, or at least, improbability, of your old beliefs in light of new knowledge and information.

But what if the question of our creation falls outside social acceptability? What would we do if we found out that the human race was the result of an experiment by a superior race of beings from another planet?

So let's take a look at what the new evidence shows, which is that some of the stories we have believed in for centuries simply cannot be true or, at best, are very misleading.

Something to keep in mind about Judaism, Christianity, and the Muslim religions is that they all hold to the absolute truth and *necessity* of Moses's being contacted by the one and only God and creator of the universe who happened to be hiding in a burning bush. Moses is said to have come down from the mountain carrying stone tablets with the Ten Commandments written on them and that he wrote the Torah according to what God told him.[94] Now just put that thought aside for a while and read on because

---

calculated the date of creation to be nightfall proceeding October 23, 4004 BC. This date is still taken quite seriously by millions of believers.

[94] The Torah is the most important document in Judaism, revered as the inspired word of God, traditionally said to have been revealed to Moses. The word "Torah" means "teaching," "instruction," "scribe," or "law" in Hebrew. It is also known as the Five Books of Moses, which are also the first five books of the Old Testament.

this turns out to be very important in today's world where about half of the planet believes this to be true to some degree.

I would like to take you on an adventure that has been ongoing for the past few decades. But first, let's take a look at Genesis, chapter 6: *The sons of God saw the daughters of man, and that they were good; and they took them for wives, all of which they chose . . .*

In older Bibles you will also find: "The Nephilim were upon the Earth in those days and thereafter too.[95] Those sons of the Gods who cohabited with the daughters of the Adam, and they bore children into them. They were the Mighty Ones of Eternity, the People of the Shem." (Genesis 6:4). In the more recent translations it has been shortened to: *There were giants upon the earth.*

Genesis 6:1-4

6:1: And it came to pass, when men began to multiply on the face of the ground, and daughters were born unto them, 2: that the sons of God saw the daughters of men that they were fair; and they took them wives of all that they chose. 3: And Jehovah said, my spirit shall not strive with man for ever, for that he also is flesh: yet shall his days be 120 years. 4: The Nephilim were in the Earth in those days, and also after that, when the sons of God came unto the daughters of men, and they bore children to them: the same were the mighty men that were of old, the men of renown.

But what if we go back even further than the Bible? What will we find? But you say, "Wait, there is no such thing as further back than the original first five books of the Bible. After all, it says, 'In the beginning.'" And you would be correct . . . if the writings in the Bible were original. But what if they are not original? We have always accepted these claims without question and based everything, and I do mean everything, on their legitimacy.

In the early 1900s, archaeologists were in Nippur, which is now Iraq, sifting through some ancient debris when they started finding fragments of clay tablets; in fact they found countless loads of clay tablets. At first they thought they were fragments of clay pots with designs made of tiny straight lines.

---

[95] "The *Book of Giants* was another literary work concerned with Enoch, widely read (after translation into the appropriate languages) in the Roman empire—. The 'giants' were believed to be the offspring of fallen angels (the Nephilim; also called Watchers) and human women." From Robert Eisman and Michael Wise, *The Dead Sea Scrolls Uncovered (1992).*

The clay artifacts they were finding were indeed quite rare and they piled the tiny clay fragments, which they considered more of an annoyance than anything, in a pile close to the dig sites. One day while working at the site they found a clay fragment that had ancient Mesopotamian writing as well as the tiny straight lines and they realized that the fragments were probably part of a vast library, a depository of some of the most ancient writings ever found.

Translating this ancient language took some time and persistence but as you may have guessed by now, it was written in what we know today as cuneiform.[96] Most of the tablets seemed to have been written by a Mesopotamian king named Ashurbanipal, who lived about 3,000 years ago. On one of the tablets Ashurbanipal says he was able to read several old languages, including a very old writing that was before the flood, which we estimate now to be around 10 thousand to 12 thousand years ago and possibly longer.

During the process of digging and finding the clay fragments, the archaeologists shipped many of the pieces all over the world for help in deciphering the unknown language. The results were not at all comforting to those in charge, who happened to be Muslims, Christians, and Jews. Even though the writings themselves were only 3,000 years old, what was being written about came from even older writings and a history that dated to well before the flood, with stories about the early formation of our planet 4.5 billion years ago and vast amounts of planetary and cosmological information. The verses seem to have been written in poetic form, probably to make them easier to memorize and pass on.

Around 5,000 to 6,000 years ago, according to the Sumerian clay tablets, there was something close to what we might call a religion. It was said that the leaders and the rulers of the people of the day were in close contact with the Gods, who were called the Anunnaki. As the translations continued it

---

[96] The cuneiform script is one of the earliest known forms of written expression. Created by the Sumerians from around 3000 BC (with predecessors reaching into the late 4th millennium Uruk IV period), cuneiform writing began as a system of pictographs. Through time, the pictorial representations became simplified and more abstract. Cuneiforms were written on clay tablets, on which symbols were drawn with a blunt reed called a stylus. The impressions left by the stylus were wedge shaped, thus giving rise to the name cuneiform ("wedge shaped"). From *www.bible-history.com.*

was shown that "Anu" meant "heaven" and "ki" meant "earth." Anunnaki literally means "those who came from heaven to earth."

An Anunnaki God

Keep in mind that the writings were written as historical fact with very little symbolism or metaphor. Those of you who may not be familiar with the works of such people as Zecharia Sitchin may be wondering why you have never heard about this before, especially since the tablets were probably the most important archaeological find in the last century and perhaps of all time. Well, there's no mystery here. You have not heard much about it because you were not supposed to. The question is: "Why" would the people in charge keep such knowledge hidden from the public?

Before we answer "why" let's look at "what" you haven't been told. The tablets include many complete stories, such as the Epic of Creation.[97] Another was Enmma Elish, which includes a description of our solar system and explains its formation. In those early days our Earth was called Tiamut. The epic tells of a great catastrophe in which a rogue planet entered our solar system, colliding with and pulverizing one planet to form what we

[97] The Epic of Creation is the Sumerian version of how the world began and includes to some degree the formation of the other members of the solar system. What's fascinating is that it parallels the Genesis story in the Bible and it predates the biblical story by many centuries and perhaps millennia.

call the asteroid belt today. Another planet, Nibiru (which was later named Marduk), was thrown into an elliptical 3,600-year orbit (its solar year was called a *shar,* which is 3,600 of our years).[98] This epic has been found at many sites throughout Iraq, so it must have been a popular story and widely known even at the time of Moses.

Another tablet called the Atra Hasis describes the Anunnaki as the inhabitants of and coming from the planet Nibiru. Apparently Nibiru was suffering some serious atmospheric degeneration and the Nibirians sent a group of explorers here to our planet in search of a metal that would help them to save their atmosphere; we refer to it as gold. By taking the gold back as a fine molecular dust, referred to as monatomical gold or white gold, and spreading it high in their atmosphere they were able to save themselves (possibly from harmful ultraviolet rays [my inclusion]). This would also explain the infatuation with and the importance placed on gold by early tribes; all they knew was the Gods valued it so it must be valuable. Another aspect to this so-called "white gold," which is gold broken down to its finest molecular parts, is that it was ingested to increase the lifespan of the Nibirians while on our planet. It was supposedly also used to make heavy objects practically weightless.

A particular genetic line of Anunnaki called the Igigi were a slightly lower form and did the actual mining in the early years. The trouble began when the Igigi began complaining about the work—that it was too demanding physically. What follows is a translation from some of the clay tablets that were found.[99]

When the Gods instead of man
Did the work, bore the loads,
The Gods' load was too great
The work too hard, the trouble too much,
The great Anunnaki made the Igigi
Carry the workload sevenfold.

---

98 It's interesting to recognize that we still have within our culture a lot of the numerical values that date to antiquity. Is it just a coincidence that the orbit of Marduk is 3,600 years and there are 360 degrees in a circle? All of this is based on a system that uses 6 and 10; this is why we have 12 constellations, 12 tribes of Israel, 24 hours in a day, 12 inches in a foot, and 12 months in a year.

99 From *The Oxford History of the Biblical World* (2001) by Michael David Coogan.

The leader of the expedition had two sons named Enki and Enlil. The Igigi met with them in a counsel meeting and the situation was fast approaching what was feared to be a mutiny. But one son, Enki, suggested a solution to this problem.

Why are we blaming them?
Their work was too hard, their trouble was too much.
Every day the earth resounded.
The warning signal was loud enough, we kept hearing the noise.
There is . . . [pieces are missing here]
Ninhursag the womb-Goddess is present
Let her create a mortal man
So that he may bear the yoke
So that he may bear the yoke, the work of Enlil,
Let man bear the load of the Gods!
[pieces are missing here]
Belet-ili the womb-Goddess is present,
Let the womb-Goddess create offspring,
And let man bear the load of the Gods!

As the story continues it is said that the female Anunnaki, Ninhursag, created a hybrid being by using what could only be genetic material from an Anunnaki and some from one of the "beasts" of the planet . . . primitive man. Written details depict many mistakes along the way, which were destroyed.[100] The Gods finally came up with a prototype they were "well pleased with" and it was called A-dam (sound familiar?), which, in their language, means something close to "duplicated and human." They even designed a smaller version for the more delicate tasks such as house servants.

Consider also the Epic of Gilgamesh (2600 BC). The Epic of Gilgamesh mentions a great flood caused by God because of man's sins. God ordered a boat to be built with one door and one window. The instructions on building the boat were given by God and the passengers on board were family members. The ark contained all kinds of animals and the end of the flood was marked by the return of a dove.

---

[100] These mistakes may have given rise to some of the mythological characters such as Cyclops, hoofed half-man half-beast beings, and so forth.

The Anunnaki beings were much larger than humans

Much of the story of Moses comes directly from the story of Sargon, who was also put adrift in a reed basket and adopted.[101] Moses was just one of a long list of "law givers"; the list includes Manou of India, Minos of Crete, and Mises of Egypt, who carried stone tablets with the laws of God written on them. Notice even the similarities in the names.

Most of the Ten Commandments are taken from the Egyptian *Book of the Dead.*

I have not stolen.	Thou shall not steal.
I have not killed.	Thou shall not kill.
I have not told lies.	Thou shall not bear false witness.

[101] The image of Sargon as a castaway set adrift on a river resembles the better-known birth narrative of Moses. Scholars such as Joseph Campbell and Otto Rank have compared the 7th century BC Sargon account with the births of other heroic figures from history and mythology, including Karna, Oedipus, Paris, Telephus, Semiramis, Perseus, Romulus, Gilgamesh, Cyrus, Jesus, and others. From www.wikipedia.com.

In fact you can find many similarities in Egyptian religion: Baptism, afterlife, final judgment, virgin birth, death and resurrection, crucifixion, ark of the covenant, circumcision, saviors, holy communion, the great flood, Easter, Christmas, Passover, and many more.

What I find truly regrettable is how knowledge such as this, which is the fact that the biblical Genesis story is an edited and copied version of older Mesopotamian texts, which were versions of even older Sumerian texts, has been completely ignored by academia and science until recently. This is another instance of the *social consequences* of religious believers controlling our information.

Now Here's My Point, and It's No Small Point

If what we've been reading in our Holy Books is nothing more than plagiarized writings of older texts, then put quite simply . . . God did not talk to Moses on the mountain, the event that gives credibility to the information found in the Old Testament. This, in essence, means the very foundation of the Judeo-Christian-Islamic religions are based on a false premise, which means that everything from that point on is false; the prophets, the stories, everything, because they all base their authority on the fact that Moses talked with God. It's no wonder they can't agree; how could they? But how do we break this news to them? I suppose we should tell them the good news first, which is that the other two religions are false; but the bad news is, *so is yours.* I am sure this information would be unsettling for religious followers but it would be the first step to get back on the path of discovery. This particular line of religious beliefs has effectively stopped our search for almost 4,000 years.

If we take into account some of the latest archaeological finds and research we could easily put the debate between creationists and evolutionists to rest because in a sense, they are both equally correct and equally incorrect. We are indeed evolutionary beings, but at the same time we did have intervention into the normal evolutionary process, hence the "missing link" and the idea that some type of intelligent designer was involved in this process. If you ever take the time to read the Epic of Creation you will soon agree that this is the same information that's in the Bible in Genesis and predates it by many centuries and probably millennia.

So, as with every other field of endeavor and research, we need to hold religious dogmas and beliefs accountable in light of new information. And in the end, we see yet another instance of where unquestioned beliefs from the Iron Age have found their way into our modern world of high technology, which as I mentioned before is a dangerous combination.

Do you see how our beliefs can keep us blind in the face of evidence to the contrary? We now have literally billions of believers on this planet who are willing to kill and die for a "story," and furthermore, the story they believe in is so distorted and edited that it lacks very little useful information. Obviously we could go into this much more deeply, but as I mentioned in the beginning of this book, there are many great books out there that you can go to and I highly recommend starting with Zecharia Sitchin's *The 12th Planet* and *Genesis Revisited* to follow up on this subject.[102] I can't overstate the importance of reading and researching the evidence and information acquired from the Sumerian Tablets.[103] There's also the Dead Sea Scrolls and the Nag Hammadi Library information to research. Of course the other option is to keep on believing, and if you're a Christian keep your eyes open to the sky for Jesus to return in his cloud, and if you're a Muslim keep a look out for Muhammad to return on his winged horse. And I don't say this to be sarcastic; this is our choice and there is simply no other way to say it. It is simply time for us, all of us, to grow up, even if it means rewriting our entire human history.

Intelligent Designer?

We have heard a lot of talk about an Intelligent Designer (ID for short) lately so a few words addressing that belief may warrant some attention. The

---

[102] Zecharia Sitchin (born 1922) is a best-selling author of books promoting the ancient astronaut theory for humankind's origins (this is not a theory in the technical scientific sense). He attributes the creation of the ancient Sumerian culture to the Anunnaki (or Nephilim) from a hypothetical planet named Nibiru in the solar system. He asserts that Sumerian mythology reflects this view. His speculations have been ignored by mainstream scientists and historians, as was Galileo and a long list of cutting-edge thinkers who went beyond the known limits of accepted doctrines.

[103] Clay tablets found in 1850 about 250 miles from Baghdad, Iraq, by an Englishman, Sir Austen Henry Layard, as he excavated the site of Nineveh, the capital of Assyria. This was near the present Iraqi town of Mosul. The tablets are among the world's oldest writing and are written in cuneiform script.

idea that we were created by an Intelligent Designer who also happens to be the creator of the whole universe brings up some difficult questions. The first thing that comes to my mind is the fact that 99.99999 percent of the universe is uninhabitable by us and indeed would instantly kill us. To think that the whole universe was created just so we would have a small piece of it to look at in the night sky is a good example of human arrogance, egocentric thinking, and naivety. Since there are more stars in the known universe than grains of sand on all the beaches on our planet, for us to be the only life-forms would be equivalent to finding one grain of sand on planet Earth that was unlike any other and the only one with microbial life. We know, at least most of us, that this is not how life works. Life is interdependent and inherently entangled with all life everywhere.

Something else to consider is the fact that more than 98 percent of everything that has ever existed is extinct. What is that about? Did the all-knowing and perfect God simply change his mind? Doesn't the Bible say he was well pleased and that it was perfect? If there is a designer he's definitely not intelligent in reference to humans, which I suppose would explain why he later decided creation was not so perfect and tried to drown us all in a great flood. No matter how you look at it, there are a lot of discrepancies and holes in this theory.

We really need to understand that the likelihood of there being other inhabited worlds much older than ours is far more likely than not and if this is the case, then we should give weight to any evidence found on this planet that would lend itself to that possibility. Of course we cannot accept this possibility blindly because we would just be creating another belief that would need defending, but if we keep analyzing the evidence we can come to some possible conclusions, and the truth is that there's more evidence every day pointing to this being the case. Of course this implies giving up some of our old beliefs and old beliefs die hard.

Say not, "I have found the truth," but rather, "I have found a truth."
Kahlil Gibran (1883-1931)

So let's move on up and take a look into the future, again by referring to old and in some cases ancient writings.

CHAPTER 40

2012—The Return of the Gods

By now many of us have heard about the year 2012 and I'm sure there are many who wonder what its significance is. Beyond the fact that there are astronomical calculations that make this date significant, we also need to consider how this information supposedly came to us. The most popular source of the 2012 information comes from the Maya, who said they received their calendar from the Gods. Consider this, the Mayan calendar is a 6,000-year calendar and is off by one day. It accurately predicts solar eclipses, the seasons, and the procession of the equinox. To come up with a calendar that exact you would need to watch and track the night skies for more than 2,000 years. Since the beings who gave this information to the Maya said they would return when the calendar runs out in 2012 it only makes sense to give it some serious attention. If this was all we knew about 2012 it would be an important date, but what a lot of people do not realize is that there are many diverse traditions that were cut off from each other that have been speaking of this date for some time. Again I want to remind you that I'm not trying to convince you, thereby creating yet another belief; I want you to see, as I have, that we live in a much more mysterious world than we were led to believe by the power figures in our lives. In fact, it may be more mysterious than we can even imagine but that should not be an excuse to make up stories to believe that keep us from searching and questioning. Here are just a few of the predictions and prophecies about the year 2012 from other traditions, peoples, and cultures:

- The Hopi predict a 25-year period of purification followed by the "End of the Fourth World" and beginning of the Fifth.
- The Maya call 2012 the "end days" or the end of time as we know it. It's also when their calendar runs out and they say that is when

the Gods will return, the same Gods who gave them the calendar, which has been shown to be one of the most accurate calendars ever found and which required millennia of astrological observations to create.

- In the Bible, Jesus says that he will be with us until the end of the "world," which is mistranslated and which actually translates to "ages" as understood in the precession of the equinox, and we are at the end of an age now.[104]
- The Maoris of New Zealand say that as the veils dissolve during this time there will be a merging of the physical and spiritual worlds.
- The Zulu tribe of South Africa believes that the whole world will be turned upside down.
- The Hindus call this the "end time of man" and that it marks the coming of Kalki and the critical mass of Enlightened Ones.
- The Incas call it the "Age of Meeting Ourselves Again."
- The Aztecs call this the "Time of the Sixth Sun," a time of transformation and creation of a new race.
- The Dogon tribe of Mali, Africa, says that the spaceship of the visitors, the Nommo, will return in the form of a blue star. They were given star maps five hundred or six hundred years ago showing the position of Sirius A and Sirius B and the length of time it took Sirius B to go around Sirius A, information we just recently were able to access through our advanced technology.
- The Pueblo acknowledge it will be the emergence into the Fifth World.
- The Cherokee ancient calendar ends exactly at 2012, as does the Mayan calendar.
- Tibetan Kalachakra teachings are prophesies left by Buddha predicting the Coming of the Golden Age.

---

[104] It's interesting to note that astronomically and mathematically it takes 2,160 years of continuous observation to know about the Earth's wobble on its axis called the precession of the equinox. You have to sit in wonder of which civilization on Earth could have accomplished this observation thousands of years ago. Of course the other possibility is that the information was given to us by an advanced civilization. Either way, it's an amazing fact that this was known in ancient times.

- According to the Great Pyramid (stone calendar) in Egypt, the present time cycle ends in the year 2012 AD.
- *"This will be a time of telepathic interplay which will eventually annihilate time."* Alice Bailey[105]
- *". . . translation or dematerialization to another sphere of the Universe."* Pierre Teilhard de Chardin[106]
- *"Our minds will unite like the fragments of a hologram."* Terence McKenna[107]
- *"As the Schumann resonance increases to 13Hz, Gaia goes from Alpha to Beta frequency and wakes up. Increasing tryptamine & beta-carboline neuro-chemistry which allows us a telepathic communion, as we become*

---

[105] Author of *The Seventh Ray: Revealer of the Age* (1995) and many other esoteric books, Alice Bailey writes books on subjects such as the path of spiritual evolution, the spiritual hierarchy, the new discipleship, spiritual meditation as a form of service for human beings, esoteric astrology, and esoteric healing. She envisages a new world religion in which every human being, regardless of race, religion, or sex, could be free to search for truth in peace and brotherhood.

[106] Author of *The Future of Man* (1959, 1964) and *The Phenomenon of Man* (1955), Pierre Teilhard de Chardin is one of the few figures in the history of the Catholic Church to achieve renown as both a scientist and a theologian. Trained as a paleontologist and ordained as a Jesuit priest, Teilhard de Chardin devoted himself to establishing the intimate, interdependent connection between science—particularly the theory of evolution—and the basic tenets of the Christian faith. At the center of his philosophy was the belief that the human species is evolving spiritually, progressing from a simple faith to higher and higher forms of consciousness, including a consciousness of God, and culminating in the ultimate understanding of humankind's place and purpose in the universe. From *The Future of Man*

[107] Author of *Food of the Gods: The Search for the Original Tree of Knowledge: A Radical History of Plants, Drugs, and Human Evolution* (1992).

used to our light bodies in preparation for the magnetic pole reversal when there will be a mass out-of-body experience."[108] Geoff Stray[109]

- *"A moment of quantum awakening. A nanosecond will be stretched into infinity and become non-time, during which we will all experience full consciousness of who we are and why we have incarnated. If we choose to return to human form, we will do so in an awakened state, as 'reflective cells' of the starmaker."* Ken Carey[110]
- *"The human race will unify as a single circuit. Solar and galactic sound transmissions will inundate the planetary field. A current charging both poles will race across the skies, connecting the polar auroras in a single brilliant flash."* Jose Arguelles[111]
- *"In 2012, Earth's axis will tilt, along with a polar reversal, causing terrestrial and celestial grids to re-align, the pineal eye will perceive beyond ultraviolet, and we'll ascend to the next dimension beyond time."* Moira Timms[112]

---

[108] Schumann resonance: Time is actually speeding up (or collapsing). For thousands of years the Schumann resonance or pulse (heartbeat) of Earth has been 7.83 cycles per second. The military has used this measurement as a very reliable reference. However, since 1980 this resonance has been slowly rising. Some measurements show it to be more than 12 cycles per second! This means there is the equivalent of less than 16 hours per day instead of the old 24 hours.

[109] Author of *Beyond 2012: Catastrophe or Ecstasy* (2009), Geoff Stray has been investigating the year 2012 for more than 20 years as a time of potentially massive significance for humankind and planet Earth.

[110] Ken Carey is the author of several books on this subject, *The Starseed Transmissions* (1991), *The Third Millennium* (1996), and *Visions* (1987), to name a few.

[111] Author of *The Mayan Factor: Path beyond Technology* (1987), visionary and historian Jose Arguelles unravels the harmonic code of the ancient Maya, providing valuable keys to understanding the next 20 years of human evolution.

[112] Author of *Beyond Prophecies and Predictions* (1996). This timely and important book synthesizes the major world prophecies—including those of the Hopi, the Maya, Babylonians, the Bible, Nostradamus, Edgar Cayce, and the Great Pyramid—into a compelling, unified theory with an inescapable message: The choices we collectively make today create our tomorrow. Positive changes in the mass consciousness and constructive actions can modify our planetary karma and avert catastrophe.

And the point is simply this; there are a multitude of ideas out there to believe in from every culture and every era in our history. Isn't it time that we start looking at all of these beliefs and start using the scientific process of inquiry to start finding answers instead of hanging on to old outdated beliefs that are dividing and destroying cultures, nations, and that are threatening to destroy our planet? A lot of different cultures have similar beliefs that converge on not only what may happen, but when. They also have astronomical evidence to back up a lot of their predictions. We should also consider the idea that if we have been visited in the past, we will be visited in the future. As it stands today, it appears that either we are being visited, as evidenced by thousands of photos and videos and eyewitness accounts, or, societies worldwide are having mass hallucinations that coincide with anomalies in our skies. This isn't even getting into the evidence released through the Freedom of Information Act through the past few years that has been collaborated by the deathbed testimonies of government officials. I would be the first to say that much, if not most, of the evidence is not true. But the evidence and testimonies that are credible are more than enough to require any thinking person to look seriously into the subject of extraterrestrial intervention. An interesting point is that a lot of people see this as credible evidence but will talk about it only in private. I'm a good example because I thought seriously about whether or not to include this information in my book for fear of ridicule. But we all need to get past this fear we have of asking the tough questions as new evidence shows up.

It is tempting to say that religious beliefs have evolved and changed through time according to our evolution, but the truth is that religious doctrines change only when the evidence gets to be so mountainous that it threatens their very existence as an institution. It's only a matter of time before most of our religious doctrines change, but do we have the luxury of time? I am deeply concerned that we do not.

CHAPTER 41

Rahasya and Alberto Villoldo, Ph.D.

New Awareness of Space and Time
A Question for Alberto Villoldo, PhD[113]

I had a conversation with Alberto Villoldo, PhD, about the shamanic way to see our world and the transformation that we seem to be headed for. He has an interesting vision into the realms where spirit and matter merge that I found enlightening. He started his career as a clinical

113 For more information visit www.thefourwinds.com.

professor investigating how energy medicine and visualization could change the chemistry of the brain. He traded it all in for a pair of hiking books and a ticket to the Amazon jungle to follow the path of a shaman. He spent more than two decades with the shamans in the jungles and high mountains of the Andes and discovered that he was more than flesh and bone; he was fashioned of spirit and light.

Rahasya: It seems to me that humanity is on several distinct layers of evolution. Some are centered in the limbic area, which values the law of the pack over individual freedom; some have become more centered in the neocortex, which values individuality and brings about things such as democracies. Yet when I look at the world today, it seems as though there is another level that's evolving, and even though it probably has the least number of people, it is by far growing the fastest. It's what is referred to by some as Samadhi, or the experience of oneness. How do you see this playing out on the world scene today with the global conflict going on?

Alberto Villoldo: Both questions you asked tie beautifully together. Our environment and our governments are in a seemingly irreversible decline. But there are still many possible futures for us, and any crisis brings with it a marvelous opportunity to envision and transform our situation. As medicine people, we must envision not the most probable future, but the best possible future for ourselves and for the planet. I see the main problem as a spiritual one, not a resource problem, or a problem with this or that government, but a larger problem centered around human beliefs, the troublesome elements founded in our mythology. Our problematic mythology is collapsing all around us. It is a mythology that is predatory, that is abusive, that reaps the cream of the earth—timber, water, topsoil—and passes the furtive costs onto future generations. These are greedy and rapacious paradigms that pose humans as a dominator over nature and over one another. And these paradigms are no longer sustainable.

Many prophecies in the indigenous world speak of this time in human history as a period of great transformation. According to the Inka Prophecy, a new day is dawning, the "Last Pachacuti," and the doorways between the worlds are opening again—a tear in the fabric of time itself.

It speaks of this time in human evolution as a chance to regain our luminous nature, to craft new physical bodies that age, heal, and die differently. Similarly, Mayan prophecies foretold of a new humanity that will be born from the ranks of the Upper World in the year 2012. What the Inka refer to as the Last Pachacuti, the Maya refer to this as the 13th Baktun. Each Baktun cycle is 394 years long and is considered its own age within "The Great Creation Cycle." The 13th Baktun represents AD 1618-2012, and it is called both "the triumph of materialism" and "the transformation of matter." The Maya predicted this final Baktun would be a time of great forgetting, when humans drift very far from their sense of oneness with nature and experience a collective amnesia of sorts. But at the closing of the 13th Baktun, they expected a great reawakening, one that would usher in a new awareness of time and space.

Now that we've taken the time to look at some opposing points of view and historical evidence that actually disputes many of our religious beliefs, let's hear what some others have to say right now, today, about our society and the process of believing without evidence.

CHAPTER 42

Lynne McTaggart

We Live and Breathe and Have Our Being in the Field A Conversation with Lynne McTaggart

Science may set limits to knowledge,
but should not set limits to imagination.
Bertrand Russell (1872-1970)

I recently had a conversation with Lynne McTaggart, who lives in London, England, and recently appeared in *Down the Rabbit Hole* (a sequel to *What the Bleep Do We Know?*). She is probably best known for her best-selling book *The Field,* and more recently, *The Intention Experiment.*[114] [115] The Intention Experiment is a global experiment to see the effects of consciousness on matter.[116] Lynne McTaggart has talked to and worked with physicists and neuroscientists from every corner of the globe.

As you read and hear what Lynne McTaggart has to say I think you will see the importance and power of belief. Used in a positive manner with validation along the way, it can be a useful tool, but used indiscriminately and backed only by faith it can lead to very undesirable consequences if perceiving reality and maintaining a sustainable environment for human existence is our goal.

I was curious to see what she would have to say about the limiting aspects of rigid belief systems. The following is a conversation I had with her on this subject and more.

Rahasya: It seems as though the science of conscious exploration includes a lot of neuroscientists. Why do you think this is?

Lynne McTaggart: There are many scientists who are biologists and physicists and they are looking into how we know what we know and how it is that we are conscious and sentient beings. There are scientists now looking at what happens to the brain when we perceive certain things and have certain beliefs. And they now have the equipment to measure what goes on in the brain, so they have been a very interesting and important resource for this kind of information.

Rahasya: I suppose with the rapid evolvement of technology it gives researchers a whole new way to research the finer energies that up to this point were entirely theoretical and out of the reach of objective research. Could you give me an example where having a rigid belief could affect the way you see reality?

Lynne McTaggart: Yes, there's a story in my book about what is going on when people are hypnotized and told something that in reality is not

---

[114] *The Field* (2008), by Lynne McTaggart.

[115] *The Intention Experiment* (2008), by Lynne McTaggart.

[116] You can be part of the Intention Experiment by logging on to www.theintentionexperiment.com.

true. In one case, they were shown pictures of Mondrian paintings which have lots of colors in them, but they were told there were no colors in them and were in black and white. They could actually measure their brain receptors for color draining away, and the receptors for black and white picking up. And so they were actually able to measure the effect of planting this conscious intention or belief in their minds that was actually contrary to the reality.

Rahasya: Yes. I remember reading in* The Holographic Universe *where a hypnotist hypnotized a father and told him, "When your daughter comes into the room I will wake you up and you will not see her."*[117] *And he not only could not see his daughter, he could actually see through his daughter to the hand of the hypnotist, indicating that our minds have the ability to phase out objects in our three-dimensional world. It's amazing when you stop to consider what our minds may be capable of. Somebody asked me the other day, "What do you find most amazing about life?" And I said, "Well, life itself is amazing, but what I really find amazing is that everybody isn't walking around completely amazed."

There are so many things we are learning about our reality that we know intellectually but no matter how many times I hear, for instance, "The world that's out there is actually in me" I'm not sure I really "get it" on a deep level.

Lynne McTaggart: Exactly. Maybe an easier way to think of it is that we are all connected, and that we are all part of something bigger, which is this giant energy field, a zero-point field that connects everything like an invisible web. And so if we think of it that way, we can see that we are really just a little part of the whole.

Rahasya: The Dalai Lama brought something up a few years ago when he asked if changing our thoughts and beliefs could physiologically change the physical structure of our brains. I think there's enough evidence now indicating that we do. What do you think the implications of this are in regard to our consciousness?

---

117 *The Holographic Universe (1992),* by Michael Talbot.

Lynne McTaggart: Well, the Dalai Lama was right. It demonstrates that the brain is a servant of our thoughts; it was some of his disciples who were tested by the University of Wisconsin's Richard Davidson, who looked at experienced meditators to see if anything was different about their brains, and found that over time their brains were lighting up the happiness portion of their brains. And the structure of their brains had become larger according to other studies. And this demonstrates that in a sense, when we continue on with great focused attention, we turn ourselves into a larger receiver and/or transmitter. There's a lot of evidence demonstrating that we are transmitting and receiving little particles of light in every moment, what they call the biophoton emission.

Rahasya: Well, this is a good segue for the next question then. It seems obvious to a lot of people on this planet that we are on the verge of taking a leap forward in our evolution in the area of spiritual consciousness. Of course this is a belief but it's a belief that scientists are validating through advanced technology. What are your thoughts on this?

Lynne McTaggart: I think that is very true and scientists are leading us there in a sense. They're leading us to a final realization, through validation, of what spiritual masters have known for centuries. That we are a lot more than we seem, that we are all part of a whole, that we have far greater extended potential than we make use of or understand. And that we have the ability now to cast off this worldview, because science tells us stories all the time, and we have been living by those stories because they are the nature of reality as it has been presented to us by modern science. For many centuries now, we have been told the Newtonian story, that things are separate, things are discrete, things are final, and that we do not have any effect on "things." We are just one more cog, one more separate element here on this very lonely planet in this lonely universe. This new evidence that quantum physics is giving us demonstrates that we are all connected and that reality resembles the kind of description given to us by many native cultures or many Eastern cultures.

Rahasya: Yes, it's amazing how many of those cultures were all but extinguished to maintain old worldviews. Of course with recent

archaeological finds those same cultures are speaking to us once again.

Lynne McTaggart: Yes, exactly.

Rahasya: I'm curious as to what your thoughts would be on this? I've noticed throughout my life that people who have been indoctrinated into what you could call rigid belief systems or ideologies seem to have their ability to conceptualize certain ideas inhibited when those ideas do not fit their preconceived views or beliefs about reality. It's not that they don't "want" to believe it; it's as if they "cannot even conceptualize" what you are talking about. This is especially true when you approach them with alternate realities and theories like string theory or spiritual consciousness. Do you think it's actually possible to inadvertently set up a neural network that would inhibit spiritual consciousness?

Lynne McTaggart: I think it's possible to set up neural networks for just about everything, this included. And I think that it would be possible if you are so convinced about something that it becomes the truth for you. And we have found that the brain cannot distinguish between a thought and a thing. And so if you keep having these kinds of limiting beliefs, you reinforce and in a sense create that which you believe to be a true. Therefore, if you are constantly inhibiting the idea of spiritual consciousness, then it would be likely you would find it almost impossible to have a spiritual awakening by setting up neural networks that aren't conducive to new ideas.

Rahasya: Yes, I have heard conflicting fundamental religious people debate, and I can tell by their dialogue that they are not listening; they are not even hearing, and I mean actually hearing, the words that the other people are using. Their minds are twisting and turning to make everything fit in their little box and it's completely impossible for them to grasp what is being presented by the "other." This obviously has huge ramifications when we think about resolving differences between people who hold beliefs that are fundamentally at odds, either one needs to be true, or the other; of course as Carl Sagan once said, "If you have two opposing beliefs and only one can

be right, that means one has to be wrong, which also means they could both be wrong."

Lynne McTaggart: Yes. That says a lot; it's their view that for them has become their truth, and their neural networks are searching to reinforce themselves in a kind of vicious circle.

Rahasya: Yes, unfortunately, vicious is where this ends up in a lot of situations throughout the ages. So what do you hope to accomplish with the Intention Experiment?

Lynne McTaggart: "Can thoughts change the world?" And if it's true, the answer will change everything.

Rahasya: Yes. I've been listening to the Beyond Belief series. It's a symposium at Salk Institute about the merging together of science and religion and what the ramifications are. Steven Weinberg was there, Sam Harris, Richard Dawkins, and many scientists and religious leaders. At one point it came up that scientists are at a disadvantage because they are bound by scientific inquiry, which demands intellectual honesty, so they cannot make a declaration that they know simply because they believe it to be true or wish it were true. They are also not as good at promoting themselves, getting the necessary funding for advertisement, and so on. So now they're in the process of trying to get more people involved, like you, like myself, to get the word out there for them, because that's not their expertise. Evangelists and fundamentalists are usually very good at getting their ideas out there in the public domain and even have global satellite networks to help.

Lynne McTaggart: We all have an important part to play in our constantly evolving consciousness. I would love to have you tell people to log on to the experiment site and get involved. This is one way to do your part and learn something along the way.

Rahasya: I have one more question that I like to ask people because it brings out some interesting comments. If you had a magic microphone allowing you to speak to everyone on the planet and you know that

whatever it was that you would say would be understood and it would have a profound effect, and you only had 30 seconds, what would you say?

Lynne McTaggart: What a great question. Okay, I think I would say the evidence demonstrates now that thoughts are not the private musings of an individual mind, that our thoughts have an effect around us. We are constantly receiving and transmitting at every moment and that this has enormous implications for everything. We need to realize that we are observers and creators, and that at every moment we are seeing our world, we are constantly remaking it at every instant, and that we have to understand too that every last thought we have, every judgment we hold, is having an effect whether it's conscious or not; the mind is sending an intention. So of course the most important thing of all is to be aware of what we are thinking and to realize it has an effect on the world around us.

You need to keep in mind that these aren't the ramblings or subjective opinions of an average person. Lynne McTaggart has talked with and studied with some of the most brilliant minds on the planet. Throughout this book we have been listening to what other brilliant minds have had to say on the subject of beliefs. What is interesting is that no matter from which direction you look at the subject of beliefs, you get the same picture and are left with the same conclusion; change is inevitable and it's happening at an accelerated rate and we either go with it or it will go on without us.

In the grand scheme of things we are not that important, and considering the fact that more than 98 percent of everything that has ever lived on this planet is now extinct, we should give this idea of change some serious thought. We are but small cogs in a great wheel connected to other even greater wheels and in the end, life would go on just fine without human participation.

CHAPTER 43

Rahasya and Drunvalo Melchizedek

The Growing Gap between Belief and Information A Talk with Drunvalo Melchizedek

My wife and I met Drunvalo Melchizedek a few years ago because we wanted to learn more about sacred geometry and its ancient roots. When you first listen to Drunvalo or read his books you can't help but wonder how he gained access to so much information. Much of what we heard seemed all but impossible at the time but little by little through the years we have watched it come to light.

In a recent conversation I had with him, the topic came up of how quickly we are learning new information. When you take the time to evaluate the speed at which we are learning you can see why the gap between old beliefs and new information is becoming unbearably wider. The reason is that, as individuals and as a society, there's a limit to how quickly we can assimilate knowledge and information. When you add to that the process of having limiting beliefs that filter out the information that comes in, it is no wonder so many people on this planet are still living with ancient superstitions and beliefs about the world we live in.

We also talked about alternative and zero point energy and many other things of interest, but I have limited this chapter to that which pertains to the subject matter of this book. You can learn much more about him and what he has to say from his website.[118] For what it may be worth, I can testify that Drunvalo did some amazing and verifiable energy work with my wife. You can also follow up with Bob Frissell, whom we also worked with and who wrote *Nothing in This Book Is True, But It's Exactly How Things Are* (1994).[119] You should also make a point of reading Drunvalo's *The Flower of Life*, Volumes I and II (1990, 2000), for a comprehensive understanding of sacred geometry and the energy fields that surround everything in existence.

Rahasya: What effect do you think it has on society living in times like ours when we are learning so much that contradicts old worldviews, whether they are religious or scientific?

Drunvalo Melchizedek: We might start out with the understanding that the earth and humanity itself is moving through a huge change or shift right now. Our whole concept and understanding of who we are and what reality is is completely different than it was 500 years ago or even 100 years ago. We have really changed. And we are moving in ways and at speeds that we cannot really understand. I mean most people can't grasp it at all, especially if they are caught in the grips of old outdated traditional beliefs and worldviews, which unfortunately most of the world is. For example, according to *Encyclopedia Britannica,* since the

---

118 www.drunvalo.net

119 Bob Frissell presents the Flower of Life workshops as presented by Drunvalo and you can find out more about him through his website www.bobfrissell.com.

beginning of recorded civilization, which was the Sumerian civilization about 6,000 years ago, until 1900, we learned a certain amount, or bits, of information; they actually counted them in a quantifiable way—basic facts about our world and reality—and it was a certain number. It took 6,000 years to be able to get to that place with that amount of information. But from 1900 to 1950, we learned as much in 50 years as we learned in 6,000 years. From 1950 to 1970, we did it again. In 20 years we learned as much as we did in 6,000 years. And it kept speeding up to about 1980. In 10 years, we learned again that same amount as we learned in 6,000 years. At this point it's going so fast that we are learning as much as we learned in 6,000 years about every month. Compiling information this quickly is bound to affect the way we view reality and our beliefs, at least for those who can comprehend what is going on. Of course beliefs that we have been indoctrinated into die hard and won't give up easy, they never have.

Rahasya: I remember Deepak Chopra saying, "Beliefs die one grave at a time." That's sad, but it's true to some extent.

Drunvalo Melchizedek: It's not only sad, it's very dangerous in a world with the technology to destroy, not only ourselves, but the planet itself. In the end it's dangerous to us because the planet will survive with or without us.

I include Drunvalo Melchizedek's remarks here because, one, Drunvalo has an incredible track record for pointing out that which seems impossible and later becomes obvious to the rest of us, and two, and this is important to the whole point of this book, he's talking about something that is verifiable and has been researched. You could also read some of Gregg Braden's work, which crosses the boundary of science and spirituality.[120] You can find his book in the bibliography.

---

[120] For more than 20 years, Gregg Braden has searched high mountain villages, remote monasteries, and forgotten texts to uncover their timeless secrets. Combining his discoveries with the best science of today, his original research crosses the traditional boundaries of science, history, and religion, offering fresh insights into ancient mysteries. In doing so he has redefined our relationship to our inner and outer worlds. From www.greggbraden.com.

CHAPTER 44

Timothy Freke and Rahasya

Waking Up from Our Dream to Lucid Living
An Interview with Timothy Freke on Gnostic Beliefs and Lucid Living[121]

[121] Timothy Freke is a popular writer who has a degree in philosophy and who has written several books on world mysticism. He has coauthored several books with Peter Gandy, including *The Jesus Mysteries* (1999), a top 10 best-seller in the United Kingdom and United States—an Amazon.com "surprise

I had the pleasure of meeting Timothy Freke and talking with him at length when he visited us in 2005 to give a lecture. One of the things I sensed about him was that he was able to maintain an open mind and heart while doing all his research on religions and religious beliefs. I bring this up because one thing I have noticed with many of the writers I have talked with or whose books I have read is that they tend to become contentious and even a little arrogant in their points of view. I noticed this in others because I watched it happening in myself while going deeper into this subject matter, which seems to take you down a rabbit hole of childish insanity very quickly.

With the coming of the information age and with people in general becoming more courageous in talking about their beliefs, it wasn't long before my eyes were opened to what has been going on in the world of organized religions and their partnership with politics and governments. When this happens to you, it's as if someone lifts a veil between you and the world you live in, and from that point on it feels strange to even have to point out the absurdity of some of the religious beliefs we hold on to. Imagine the frustrations you would feel in talking with a grown adult and explaining to him or her that there is no Easter Bunny; even the need to explain it would feel a little ridiculous.

Somehow, Timothy has maintained an open heart when communicating with people on this subject and I applaud him for that because this, in the final hour, is what it takes to get a person to understand the deeper meanings of what is being said. I highly encourage you to buy his little book (you

---

best-seller"—and *Jesus and the Lost Goddess* (2007), an exegesis of Christian (especially Gnostic) mythology with regard to interpretation and philosophy. Recently he was interviewed for the History Channel documentary *Beyond The Da Vinci Code* and has been the focus of *Modern Mystics* on the BBC World Service and *Beyond Belief* on BBC Radio. Freke's recent work, *The Laughing Jesus: Religious Lies and Gnostic Wisdom* (2005), again coauthored with Peter Gandy, is a critique of religious literalism and affirms mystical spirituality. This book follows from *The Jesus Mysteries,* which is an examination of the evidence for a historical Jesus and parallels Paganism with early Christianity. Although backed up by scholarly research, Freke and Gandy's works are written to be accessible for a popular audience. Freke has written on Gnosticism, the ancient religious movement. However, Freke also uses the term "Gnosticism" to describe his mystical approach, contrasted with "Literalism," which he defines as rigid dogmatic religious belief and criticizes as a dangerous force in today's world.

can read it in an hour) called *Lucid Living* (2005); it will, as he states, turn your world inside out.

His book *The Laughing Jesus: Religious Lies and Gnostic Wisdom* (2005) is divided into two sections; the first is called "The Bathwater" and the second is called "The Baby." One reason this is so important is that we have a tendency to become polarized in our thoughts and beliefs; something is either right or wrong. When this happens we tend to throw out the whole idea, concept, philosophy, or belief. Many wise and profound thoughts are to be found throughout the Bible and Quran but unfortunately they are mixed in with so many ideas that are so completely wrong and immoral that an average person living in today's world could never live by the laws and moral codes they set forth, although there are groups that are trying their best to do so and it has driven them into a rage of madness and polarized them against the rest of the civilized world.

As I mentioned earlier, one of Joseph Campbell's greatest fears when it came to religion was that the day might come when science and reason would undermine organized religion to the point where we would throw the whole idea of religion and God out the window. His thoughts were that within religious stories there are important mythological ideas and truths that are easily transferred from one generation to another without being dependent on the colloquialisms of the day to understand. The use of myth helps to maintain the purity of the message and the depth to which it can be understood.

A Conversation with Timothy Freke

Rahasya: How important do you feel the knowledge about Lucid Living, waking up from our dreams and beliefs, is in today's world and why?

Timothy Freke: I think this is a perennial wisdom, which has been relevant in every age. It's not just something that refers to today and it's not something that's stuck in the past. It's about the human condition.

This perennial wisdom reveals that life is a journey of becoming more conscious and coming to know who we really are. In other words, life is a journey from unconscious oneness to a state of conscious oneness. We start as members of the unconscious herd and have the opportunity to become more conscious, through doubting and thinking for ourselves, until we enter the ultraconscious state I call being "deep awake." Then we see that in reality we are all one. We see that in reality there is one primal awareness dreaming this great dream of life, and appearing to itself in all these different forms.

The reason this understanding is so relevant to today's world is because it reveals who we really are and that essentially we are all one. It shows us that life is a journey to recognize our own true nature. And I can't think of anything more important than that.

Rahasya: That's true, especially since so many of our old paradigms, beliefs, and mythologies are failing us in today's world. What was it in your life that made you start questioning the reality of how we view life? Was there a particular instance?

Timothy Freke: Yes, there was. It started for me when I was about 12. I always felt that the grown-up world was in some sort of collective denial. It seemed to me that no one really knew what the hell was going on, but just not admitting it. Everyone was pretending they understood what life was all about and acting as if they knew it made sense, but actually it didn't. So I used to just sit on this hill above the sleepy little town in southwest England I grew up in, with my dog, just thinking about things . . .

And then one day for absolutely no reason that I can put my finger on, I had my first experience of being "deep awake" and living lucidly. It was like waking up from a dream. Suddenly I knew that all is one. My memory is of an experience of overwhelming love. It was if I was being held in this universal love and as if I *was* this love. It's truly beyond words to describe.

I don't know how long the experience lasted, probably not very long. And when I came down from the hill, I went back to being a confused teenager. But something had changed. I had this seed planted, and from then on my life was dominated by wanting to know what the hell had happened to me and wanting to get back to that experience. That's what made me seek out people who seemed to have had similar experiences.

Since then I've found you can approach this experience in many, many ways and it's available at all times.

In some ways what I think I've done with my little *Lucid Living* book is written the book that I would have loved to have read when I was 12. It would have made sense of that experience for me in the simplest way. Rather than having to go through this journey I've been on, through all these different spiritual traditions and different philosophies, seeking for an understanding that is really very simple, so I could say, "Ah . . . now I see."

***Rahasya: I see what happened to you at the age of 12 as one of the best clues ever given to a seeker because if you can grasp that transition from dreaming to awakening you can grasp this one from awakening from our everyday life to enlightenment. I believe it was Hermes that said,* "As it is above, so it is below."**

Timothy Freke: Yes, exactly.

Rahasya: When you talk about Lucid Living, are you referring to what some call global or planetary awakening?

Timothy Freke: I think we are waking up to oneness in many ways. We have a growing recognition that we are on one planet and that there is an ecological crisis which affects the whole globe. There is a growing recognition that we are one species, and now we have much more compassion for people who just happen to be from different races or tribes. We are talking about things in new ways that were unheard of in earlier times and that is a huge step forward. There is a feeling that we could wake up as a planet and I feel this is a strong possibility.

Certainly I think we are growing, we are waking up from our limiting beliefs. This is something that each individual has to experience for themselves. But, paradoxically, when we wake up as an individual, we see that we are essentially one with the whole universe.

When we become "deep awake" and live lucidly we see that each one of us has two poles to our identity. At one pole each of us is a unique perspective on time and space, which we call a person. And at the other pole there is our shared deeper nature, which is the primal oneness from which everything is arising. When we are conscious of both poles of our identity we see that the multiplicity of life is arising from the essential oneness. When we are conscious of our essential nature we can meet in the oneness as well as in the separateness. And that is an experience of unconditional compassion I call "big love."

Rahasya: When you look at the complexity of the world we live in, and then add to that our human arrogance, nationalism, deteriorating environments, toxic by-products of consumerism, and the division caused by our various beliefs, it seems that we have created a world in which there are no solutions and is obviously unsustainable. Albert Einstein once said,* "The significant problems we face cannot be solved at the same level of thinking we were in when we first created them." *In other words, most of our problems today were created by a consciousness and a belief in separateness, so do you see Lucid Living as a level of awareness that would make it possible to come up with lasting solutions to today's problems?

Timothy Freke: That is a fantastic question. When I look at the world's problems, such as environmental degradation, AIDS, war, poverty, prejudice, big business and all that, I feel it's very easy to become completely overwhelmed. It's just too much. It's too much for us to face and it makes us pessimistic and I think this is why there is a general pessimism around. There was a big burst of optimism in the '60s when people felt we were going to change the way we live, especially in the West, and it's been followed by a kind of pessimistic realism, which I don't think is realistic at all.

The problem is that we are caught up in a story which makes us feel cynical and unable to take the necessary steps to continue this journey. Yet I see that we could change the world. You know, Archimedes said, *"Give me a lever and I'll move the world."* And I've wondered for a long time, "Is there a lever that we could move the world with? Is there a simple idea, a big idea that could move us on?" And I now feel there is. And that big idea is the big idea that's been around for centuries, which is that we are one.

All of our suffering, collectively and individually, ultimately arises from one simple mistake. We are convinced we are separate, but in reality we are not. This is why the realization of our essential oneness can change everything. Because if we come to recognize we are one with everything and everyone, then we'll find ourselves in love with everything and everyone. And then we'll find creative and compassionate solutions to all our problems.

I don't doubt for a moment that our imagination, which is the imagination of the universe, can come up with solutions to all of the problems we face. If we recognize that we are not just these little separate individuals that we believe ourselves to be, and we recognize we are one and live with that understanding, then I think we can overcome any challenge. And that makes me more optimistic; because if I've got to take on all these huge world problems, it's more than I can cope with. But if I think to myself, all I need to do is spread one idea; that seems entirely practical.

If people can really get the idea of oneness then they'll start to see through the veneer of separateness which divides us. It could be like when the Berlin Wall unexpectedly came down. We could be surprised by how easily and suddenly we entered a new level of consciousness. If we can just bring this mystical wisdom, which men and women have been articulating in every time and every culture, into mainstream society . . . if we could really grasp it collectively, it would change our consciousness and that would change the world. And then we could make this dream of life the joyous celebration we all want it to be.

***Rahasya: As Victor Hugo said,* "There is only one thing more powerful than all the armies of the world and that is an idea whose time has come."**

Timothy Freke: I think so too. And I hope the time has finally come for us to collectively embrace the idea of awakening to oneness.

FINAL THOUGHTS

As I type out the words "Final Thoughts" I can't help but wonder if some of us have reached a point where we could have any final thoughts about anything. We are so accustomed to having our thoughts given to us as beliefs that it is difficult for some of us to do otherwise. I hope in some small way this book has brought you, at least, a partial remedy by helping you see the underlying mental patterns that have us trapped in illusion. But I am a realist and, as such, realize that many people will not read this book and some of those who do will miss some of the points being made. It is all too easy to get stuck in our emotions and belief patterns, fighting desperately to maintain the status quo.

Even as I am finishing this book in October 2009, the research coming back from fMRIs is validating everything in this book and then some. For instance, it appears that whenever we are given a thought or idea to believe, it normally goes to certain areas of the brain where it is processed and held in memory. The problem arises when the thought or idea is so absurd to the mind that it cannot reconcile it with the reality that it experiences. At that point the mind sends it to the part of the brain that normally processes emotions because that part of our brains does not need to validate information; it "just feels it to be true." This is where we get ourselves into problems because we "feel" we are right so it must be true.

Then there's the chemistry around feeling and even "knowing." It has been clearly shown for many years that the sense of "knowing something is true" can be induced chemically. We are far from the final word on this but it's clear that we must start having open and honest enquiry into the nature of our thoughts, emotions, and beliefs. Many of the beliefs we have fought and died for are now topics discussed in classrooms on how lost we have been in our past. What might we be doing and believing today that will be discussed in future classrooms?

I would also like to add this one last point. A lot of the information you have read in this book could be taken to be negative. One thing I have learned in life is that it is crucial to know what you are up against to release it, because to stay attached to it either mentally or emotionally can be damaging in many ways. So now it's time to think positively, and for this reason there will be no second edition or a follow-up book because I too will be letting this go and any future books will be the positive approach to this subject, which is the mysterious adventure we refer to as our lives.

So now it's out of my hands and in yours, so ask yourself, "What now?"

- Get involved. Thousands of organizations worldwide are doing everything in their power to bring more awareness to this planet and are living examples of what happens when spirit takes action.
- Join environmental groups that are fighting back against corporations and politicians that could not care less about the next 20 years of life on Earth, let alone the next 100 years.
- Don't blindly believe anyone. Check things out for yourself, ask questions, stand up for yourself because no one else is going to do it for you. If enough of us do this we will turn things around. But we need to do it quickly while we still have a democracy that has the machinery to fix itself.
- Write a "living will" to make sure your money and assets go to worthy causes and family members who will do the right thing.
- Think green in everything you do. Walk to work, stop using paper cups, stop buying new cars and look for good used ones, support alternative energy and businesses that use alternative energy. One reason automotive industries are having a difficult time is that they are part of a system that is dying.
- Do everything you can to become more present in your thoughts, meditate, communicate with others about things that you have read in this and other books or watched on the Internet. As Ram Dass once said, *"Be Here Now,"* which is a deceptively simple statement but all three words are profound.
- Stop buying more stuff that ends up in the dump, all in the name of making you feel good for a minute or two. We are running out of resources. We are at the end of an exponential population explosion that threatens our very survival and as you know by now, the chances of a savior's showing up at the last minute and rescuing us is remote at best.

- Reach out to your circle of influence, your friends and family, community, and then to your state government, and make yourself heard through numbers. This has a profound effect on grants that are given and not given out to historians, archaeologists, scientists, biologists, researchers, and so on.
- Even though our government has managed to make the voting system all but dysfunctional, it does still work to some degree, so use it. And keep in mind that every time you buy something, you are voting, so shop consciously and wisely.
- Remember that you are a great spirit, powerful beyond human comprehension. In the end, there are no "them" and "us." We all created the world we are living in. That's the bad news and the good, good because that means we can create another world, *A New Earth,* as Eckhart Tolle would say.

It's really not that difficult to figure out where to start and what to do. The difficult part is getting out there and doing it. I will be regularly updating my website for updates on religion and science and the latest research in neuroscience. I also have a section on each person I have interviewed with video and more information about him or her and complete interviews on my website at www.RahasyaPoe.com; email: rahasya@usa.com.

In closing I would like to thank you for sharing your time with me to explore new ideas in the hope of cocreating a better world.

Rahasya Poe

BIBLIOGRAPHY

Abbott, Jennifer. *The Corporation* (documentary DVD). 2004.

Ali, Maulana Muhammad. *The Holy Qur'an with English Translation and Commentary*. Ahmadiyya Anjuman Ishaat. 1991.

Becker, Ernest. *Escape from Evil.* Free Press. 1975.

Becker, Ernest. *Revolution in Psychiatry: The New Understanding of Man.* Free Press. 1964.

Becker, Ernest. *The Denial of Death.* Collier-Mac. 1973.

Begley, Sharon. *Train Your Mind, Change Your Brain: How a New Science Reveals Our Extraordinary Potential to Transform Ourselves.* Ballantine Books. 2007.

Bose, Mandakranta. *Ramayana Revisited.* Oxford University Press. 2004.

Boyer, Pascal. *Religion Explained: The Evolutionary Origins of Religious Thought.* Basic Books. 2001.

Braden, Gregg. *Walking between the Worlds: The Science of Compassion.* Radio Bookstore Press. 1997.

Dalai Lama. *The Universe in a Single Atom: The Convergence of Science and Spirituality*. Broadway. 2006.

Davis, Gregory. *Islam: What the West Needs to Know* (documentary DVD). Quixotic Media. 2009.

Dawkins, Richard. *The God Delusion*. Mariner Books. 2006.

Dennett, D.C. *Breaking the Spell: Religion as a Natural Phenomenon*. Viking. 2006.

Dispenza, Joseph. *Evolve Your Brain: The Science of Changing Your Mind.* HCI. 2007.

Dispenza, Joseph. *God on Your Own: Finding a Spiritual Path outside Religion.* Jossey-Bass. 2006.

Downing, Barry. *The Bible and Flying Saucers.* Marlowe & Company, 1968, 1997.

Ehrman, Bart D. *Misquoting Jesus: The Story behind Who Changed the Bible and Why.* Harper San Francisco. 2005.

Freer, Neil. *Breaking the Godspell: The Politics of Our Evolution*. The Book Tree. 2003.

Freke, Timothy. *Lucid Living*. 2005.

Freke, Timothy, and Peter Gandy. *The Jesus Mysteries: Was the Original Jesus a Pagan God?* Three Rivers Press. 1999.

Freke, Timothy, and Peter Gandy. *The Laughing Jesus: Religious Lies and Gnostic Wisdom*. Harmony Books. 2005.

Furman, Mark Evan, and Fred P. Gallo. *The Neurophysics of Human Behavior: Explorations at the Interface of Brain, Mind, Behavior, and Information*. CRC Press. 2000.

Gardner, Laurence. *Bloodline of the Holy Grail*. Barnes and Noble Books. 1997.

Gardner, Laurence. *Genesis of the Grail Kings*. Fair Winds Press. 2002.

Hardt, James, PhD. *The Art of Smart Thinking*. Biocybernaut Press. 2007.

Harris, Sam. *Letter to a Christian Nation*. Vintage. 2006.

Harris, Sam. *The End of Faith: Religion, Terror, and the Future of Reason*. W.W. Norton. 2005.

Hawking, Steven. *A Brief History of Time*. Bantam. 1988.

Hirschfeld, Lawrence A., and Susan A. Gelman (eds). *Mapping the Mind: Domain Specificity in Cognition and Culture*. Cambridge University Press. 1994.

Hitchens, Christopher. *God Is Not Great: How Religion Poisons Everything*. Twelve Hatchet Book Group. 2007.

Jacoby, Oren. *Constantine's Sword* (documentary DVD). Storyville Films. 2007. An excellent documentary based on the James Carroll's exploration of the outcome of religious intervention into governments over the past 17 centuries.

Jarecki, Eugene. *Why We Fight* (documentary DVD). Sony Pictures. 2005.

Jaynes, J. *The Origin of Consciousness and the Breakdown of the Bicameral Mind*. Houghton Mifflin. 1976.

Jones, Marie D., and Larry Flaxman. *The Resonance Key: Exploring the Links between Vibration, Consciousness, and the Zero Point Grid*. New Page Books. 2009.

Kenyon, J. Douglas. *Forbidden Religion: Suppressed Heresies of the West*. Bear & Company. 2006.

Leedom, Tim C., & Maria Muroy. *The Book Your Church Doesn't Want You to Read*. Cambridge House Press. 2007.

Lipton, Bruce L. *The Biology of Belief: Unleashing the Power of Consciousness, Matter, and Miracles.* Hay House. 2005.

McTaggart, Lynne. *The Field: The Quest for the Secret Force of the Universe.* Harper Paperbacks. 2003.

McTaggart, Lynne. *The Intention Experiment: Using Your Thoughts to Change Your Life and the World.* Free Press. 2007.

Melchizedek, Drunvalo. *The Ancient Secret of the Flower of Life, Vols. 1 & 2.* Light Technology Publishing. 2000.

Murdock, D.M. *Who Was Jesus? Fingerprints of Christ.* Steller House. 2007.

Newberg, Andrew, MD. *Why We Believe What We Believe.* Free Press. 2006.

Nicoll, Maurice. *The New Man: An Interpretation of Some Parables and Miracles of Christ.* Shambala Publications. 1987.

Osho. *Hidden Mysteries.* The Rebel Publishing House. 1997.

Osho. *The Mustard Seed: The Gnostic Teachings of Jesus the Mystic.* Element. 2004.

Ouspensky, P.D. *A New Model of the Universe.* Vintage Books. 1997.

Ouspensky, P.D. *The Fourth Way.* Alfred A. Knopf. 1957.

Ouspensky, P.D. *The Psychology of Man's Possible Evolution.* Vintage Books. 1973.

Pearson, Bishop Carlton. *The Gospel of Inclusion: Reaching beyond Religious Fundamentalism to the True Love of God.* Azusa Press International. 2006.

Pert, Candace B., PhD. *Molecules of Emotion: The Science behind Mind-Body Medicine.* Scribner. 1999.

Radin, Dean. *Entangled Minds.* Pocket Books. 2005.

Rajneesh, Bhagwan Shree. *Zarathustra, the Laughing Prophet.* Rebel Publishing House. 1987.

Ripper, Velcrow. *Fierce Light: When Spirit Meets Action* (documentary DVD). 2009.

Russell, Bertrand. *Why I'm Not a Christian.* Routledge. 1957.

Russell, Peter. *From Science to God: A Physicist's Journey into the Mysteries of Consciousness.* New World Library. 2005.

S, Acharya. *Zeitgeist* (DVD). Produced by Peter Joseph. 2007. Both DVD and *The Companion Guide* can be downloaded from the website www.zeitgeistmovie.com.

Sagan, Carl. *The Demon-Haunted World: Science as a Candle in the Dark.* Ballantine Books. 1997.

Sagan, Carl. *The Pale Blue Dot.* Headline. 1995.

Saranam, Sankara. *God without Religion: Questioning Centuries of Accepted Truths*. Pranayama Institute. 2007.

Schwartz, Jeffrey M., and Sharon Begley. *The Mind and the Brain: Neuroplasticity and the Power of Mental Force*. Harper-Perennial. 2003.

Shen, Patrick. *Flight from Death: Quest for Immortality* (documentary DVD). 2003. www.flightfromdeath.com.

Shermer, Michael. *How We Believe: Science, Skepticism, and the Search for God.* Owl Books. 1999.

Shermer, Michael. *Science Friction: Where the Known Meets the Unknown.* Owl Books. 1995.

Shermer, Michael. *Why Darwin Matters: The Case against Intelligent Design.* Owl Books. 2006.

Sitchin, Zecharia. *Genesis Revisited.* Bear & Company. 2002.

Sitchin, Zecharia. *The 12th Planet*. Avon Books. 1990.

Sitchin, Zecharia. *The Earth Chronicles Expeditions: Journeys to the Mythical Past.* Bear & Company. 2004.

Smith, John D. *Mahabharata*. Penguin Classics. 2009.

Talbot, Michael. *The Holographic Universe.* Harper Collins. 1992.

Tarico, Valerie, PhD. *The Dark Side: How Evangelical Teachings Corrupt Love and Truth.* Dea Press. 2006.

von Daniken, Erich. *History Is Wrong*. Career Press. 2009.

Wallace, B. Alan. *Contemplative Science: Where Buddhism and Neuroscience Converge*. (Columbia Series in Science and Religion). Columbia University Press. 2006.

Weinberg, Steven. *Dreams of a Final Theory*. Vintage. 1993.

Weinberg, Steven. *The First Three Minutes: A Modern View of the Origin of the Universe.* Basic Books. 1988.

Wolpert, Lewis. *Six Impossible Things before Breakfast: The Evolutionary Origins of Belief.* W.W. Norton & Company. 2007.

Yogananda, Paramahansa. *God Talks with Arjuna: The Bhagavad Gita*. Self-Realization Fellowship. 2001.

Caught in Make-Believe

Life is easy the way we try;
Believe it all and never ask why.
I know it's hard to give this thought;
It's make-believe and you know you're caught.
One man believes what another man knows;
But Faith in Truth is true when it flows.
We stand on faith when we have no ground;
The fear to lose it is we might fall down.
But what's its use? To fall we should,
For if it's not true, then what is its good?
So Faith is just another word you use;
To protect your lies, that you fear you'll lose.
And if you're not careful, and refuse to see;
You'll lose yourself, and start a war with me.

Wm. "Rahasya" Poe - 1974

About the Author

Living in a world where the options for creating a sustainable society are quickly disappearing, Rahasya Poe has spent many years questioning some of our most cherished and established answers that no longer serve us and are in fact dysfunctional and dangerous in today's world. His journey has taken him from practicing meditation in India to the ruins of Peru to sitting with correnderos in Mexico and taking Ayuhaska in Brazil. He now publishes a magazine in northern California with his wife Dhara.

Get Published, Inc!
Thorofare, NJ 08086
18 November 2009
BA2009261